an electronic companion to
statistics

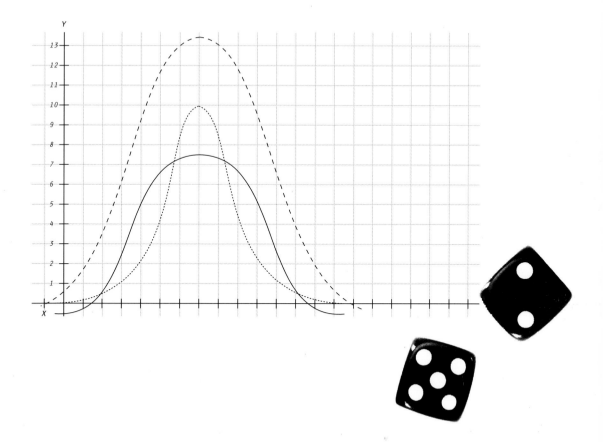

an electronic companion to
statistics™

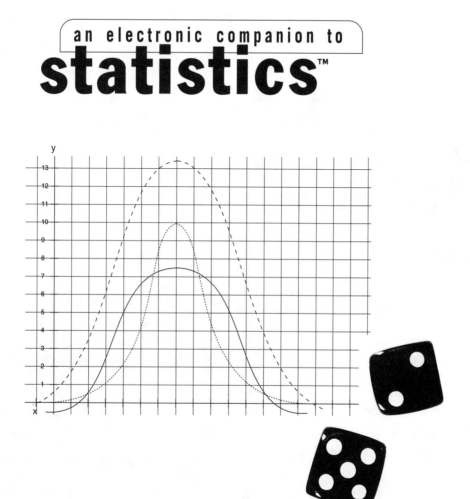

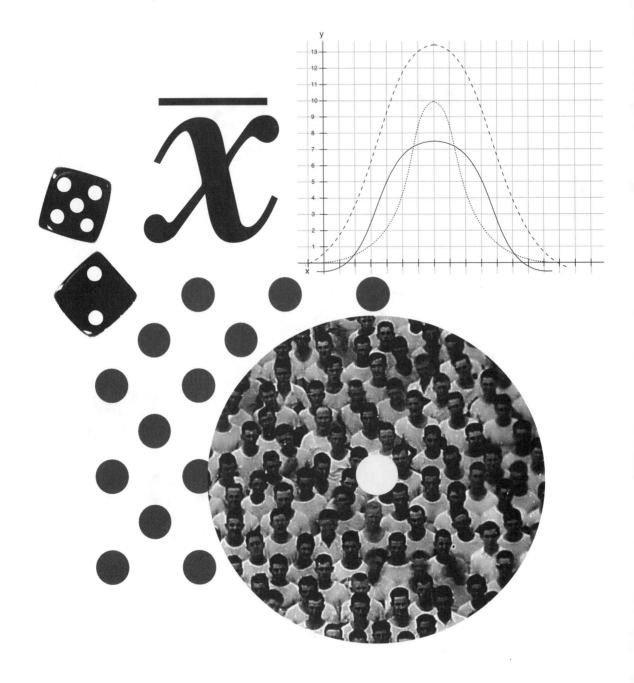

an electronic companion to
statistics™

George W. Cobb
Mathematics, Statistics, and Computer Science
Mount Holyoke College

and

Jeffrey A. Witmer *Oberlin College*
Jonathan D. Cryer *University of Iowa*

with the assistance of
Peter L. Renz
Kristopher Jennings

COGITO

Cogito Learning Media, Inc.
New York   San Francisco

© 1997, Cogito Learning Media, Inc.

ISBN: 1-888902-40-X

COGITO END-USER AGREEMENT

PLEASE READ THE FOLLOWING CAREFULLY BEFORE USING THIS PRODUCT (INCLUDING THIS WORKBOOK AND THE ACCOMPANYING SOFTWARE AND OTHER MATERIALS) (THE "PRODUCT"). BY USING THE PRODUCT, YOU ARE AGREEING TO ACCEPT THE TERMS AND CONDITIONS OF THIS AGREEMENT. IF YOU DO NOT ACCEPT THOSE TERMS AND CONDITIONS, please return the Product (including any copies) to the place of purchase within 15 days of purchase for a full refund.

LIMITATIONS ON USE

Cogito Learning Media, Inc. ("Cogito") licenses the Product to you only for use on a single computer with a single CPU. You may use the Product only on a stand-alone basis, such that the Product and the user interface and functions of the Product are accessible only to a person physically present at the location of the computer on which the Product is loaded. You may not lease the Product, display or perform the Product publicly, or allow the Product to be accessed remotely or transmitted through any network or communication link. You own the CD-ROM or other media on which the Product is recorded, but Cogito and its licensors retain all title to and ownership of the Product and reserve all rights not expressly granted to you.

You may not copy all or any portion of the Product, except as an essential step in the use of the Product as expressly authorized in the documentation included in the Product. You may transfer your license to the Product, provided that (a) you transfer all portions of the Product (including any copies), and (b) the transferee reads and agrees to be bound by the terms and conditions of this agreement. Except to the extent expressly permitted by the laws of the jurisdiction where you are located, you may not decompile, disassemble or otherwise reverse engineer the Product.

LIMITED WARRANTY

Cogito warrants that, for a period of 90 days after purchase by you (or such other period as may be expressly required by applicable law) ("Warranty Period"), (a) the Product will provide substantially the functionality described in the documentation included in the Product, if operated as specified in that documentation, and (b) the CD-ROM or other media on which the Product is recorded will be free from defects in materials and workmanship. Your sole remedy, and Cogito's sole obligation, for breach of the foregoing warranties is for Cogito to provide you with a replacement copy of the Product or, at Cogito's option, for Cogito to refund the amount paid for the Product.

EXCEPT FOR THE FOREGOING, THE PRODUCT IS PROVIDED WITHOUT WARRANTIES OF ANY KIND, EXPRESS OR IMPLIED, INCLUDING, BUT NOT LIMITED TO, ANY IMPLIED WARRANTIES OF MERCHANTABILITY OR FITNESS FOR A PARTICULAR PURPOSE. Among other things, Cogito does not warrant that the functions contained in the Product will meet your requirements, or that operation of the Product or information contained in the Product will be error-free. This agreement gives you specific legal rights, and you may also have other rights which vary from state to state. Some states do not allow the exclusion of implied warranties, so the above exclusion may not apply to you. Any implied warranties will be limited to the Warranty Period, except that some states do not allow limitations on how long an implied warranty lasts, so this limitation may not apply to you.

LIMITATION OF LIABILITY

In no event will Cogito be liable for any indirect, incidental, special or consequential damages or for any lost profits, lost savings, lost revenues or lost data arising from or relating to the Product, even if Cogito has been advised of the possibility of such damages. In no event will Cogito's liability to you or any other person exceed the amount paid by you for the Product regardless of the form of the claim. Some states do not allow the exclusion or limitation of incidental or consequential damages, so the above limitation or exclusion may not apply to you. Also, this limitation will not apply to liability for death or personal injury to the extent applicable law prohibits such limitation.

GENERAL

This agreement is governed by the laws of the State of California without reference to its choice-of-law rules. This agreement is the entire agreement between you and Cogito and supersedes any other understandings or agreements. If any provision of this agreement is deemed invalid or unenforceable by any court or government agency having jurisdiction, that particular provision will be deemed modified to the extent necessary to make the provision valid and enforceable, and the remaining provisions will remain in full force and effect. Should you have any questions regarding the Product or this agreement, please contact Cogito at 1-800-WE-THINK.

2 3 4 5 6 7 8 9 10—RRD—00 99 98 97 Printed in U.S.A.

contents

Preface

FOUR KINDS OF UNDERSTANDING

To gain a thorough understanding of statistics, you will need to combine several different kinds of thinking:

1. Computational/numerical;
2. Visual/graphical;
3. Verbal/interpretive;
4. Structural/deductive.

Computational/numerical. This is what many people think of when they think of statistics: memorizing rules and formulas, then plugging in numbers and doing a lot of calculating. Although statistical work does rely on a lot of computing, there is almost no value in just memorizing rules. If you find yourself doing that, you are probably not using your time efficiently. Learning the rules can be a useful step toward understanding, however, *provided you relate them to their visual, verbal, and structural meanings.*

Visual/graphical. Learning how to use graphs to represent numbers and how to read patterns in graphs is essential to understanding statistics. The mechanical part—constructing the graph—is just the beginning. The more important part is learning to *think* visually about numbers.

Verbal/interpretive. "Data are not just numbers, but numbers with a context."[1] In statistics, any calculating you do, and any graphs you construct, ought to be part of a search for meaningful patterns in real data. (Exception: When you're *first* learning a new skill, it is sometimes useful to practice with lists of numbers that have no context, but only at first.) Try to make it a habit to ask, "What does this tell me about the data (in relation to its context)?" The search for meaning is what makes statistics worthwhile.

Structural/deductive. The methods and concepts of statistics are related to each other by a logical structure. This structure is part of the "big picture" that should gradually come into focus for you as you work at understanding statistics.

---

[1]Moore, David S. (1992). "Teaching Statistics as a Respectable Subject," in Florence Gordon and Sheldon Gordon, eds., *Statistics for the Twenty-First Century*, MAA Notes no. 26, Washington: The Mathematical Association of America.

FOUR KEY THEMES

One approach to the big picture is to relate what you learn to four basic themes.

1. Production;
2. Exploration;
3. Repetition;
4. Inference.

Production How and why were the data produced? This may seem like an obvious question to ask, but as a rule, it doesn't get enough attention. Bad planning and careless data collection can ruin an experiment or survey. Principles of good planning and careful collection can't be put into mathematical formulas, but the principles are important all the same.

Exploration Data = Pattern + Deviation. Data analysis is intended to be a search for meaningful patterns. No one pattern is likely to tell the whole story, though. A useful pattern must be simpler than the actual data, which means that the data deviate from the pattern, at least a little. Exploring a data set means trying out a variety of patterns to see how well they fit: What's the balance between pattern and deviation?

Repetition What will happen if I repeat this a large number of times? Statistics is sometimes defined as the science of learning in the presence of *variation*. If I do an experiment today and repeat the same experiment next week, the two sets of results will most likely be somewhat different. Statistics has value because it gives us a way to learn from the results of a single experiment or survey, even though we would get somewhat different results from a second one. Imagine repeating the experiment a large number of times and looking at all the results together. Part of what we see will be pattern—those aspects of the results that we expect to be the same from one repetition to the next. The rest will be deviation—the part that varies from one repetition to the next. We can sometimes use statistical logic (inference) to decide, from the results of just one experiment or survey, what patterns to expect from a large number of repetitions.

Inference Be suspicious of any theory that makes your data into an unlikely outcome. One major branch of statistics deals with production: how to plan experiments and surveys. Another deals with exploration: finding and describing patterns in data.

Yet another deals with inference: drawing conclusions about the long-run patterns we think would emerge if we were to repeat the data-production process a large number of times. For any set of results, we can try out a variety of theories (usually called models and hypotheses) about the process that created our data. We can then ask what sort of data our theoretical process (model, hypothesis) would be likely to generate. The basic logic of inference is that we should not ordinarily trust any model that would be unlikely to give results like our actual data.

This logic takes some getting used to, not because it is unfamiliar, but because in everyday life it is so automatic we don't think about it. (For example, you automatically rule out rain when the sky is cloudless.) Statistics uses the same logic in a more formal way.

This introduction is necessarily somewhat abstract. The four kinds of understanding and four basic themes will take on more meaning as you have the chance to relate them to the methods and concepts of statistical thinking.

MANY KINDS OF PROBLEMS

This workbook has problems of various kinds.

Simple drills for basic skills. These problems use simple, made-up data, to be easy and quick. Their purpose is to help you learn the mechanical skills. Just as a piano player practices scales and a beginning language student learns rules of grammar, you may find it useful, at first, to practice the mechanics in a simple setting with no "story" to take your mind away from learning the skills.

Skills in context. Just as pianists practice scales in order to build skills for playing real music, and language students learn grammar in order to read literature or have real conversations, your goal in learning statistical skills should be to use them to find meaning in real data. Many of the problems in the workbook could be labeled "skills in context"; they ask you to apply your skills in real-world settings with authentic data. At this stage in your learning, you may find it helps to have some coaching about what to look for. The problems are designed with that in mind.

Simple drills on basic concepts. Along with the numerical skills, you also need to practice the basic concepts you need for statistical thinking. We've included a number of problems that ask you about these ideas in a "clean" setting with no "story."

Concepts in context. Once you have the basic ideas clearly in mind, you'll be ready to go on to the harder but more rewarding problems that ask you to use the ideas of statistics in a real-world setting.

ACKNOWLEDGMENTS

My thinking about how best to help others learn statistics, and about statistics itself, owes much to many. In particular, some of the ideas, phrases, and metaphors of my colleagues are so good that I can't imagine a more effective alternative. It is a pleasure to acknowledge these brilliant gems, and the sparkling minds that brought them to us.

David S. Moore, of Purdue University, is the author of the four-word outline-cum-mantra for describing distributions ("Plot, shape, center, spread"), along with its cousin ("Plot, shape, direction, strength") for scatter plots. He is also the one who urges us all to present probability distributions (generally) and sampling distributions (in particular) as answers to the simple question, "What would happen if I were to repeat this many times?" Finally, he is the creator of the prize-worthy Madison Avenue sound bite "less variable, more normal" for summarizing the effect of increasing sample size on the sampling distribution of the mean. To those of us who have thought long and hard about statistics, David's ability to compress the sometimes swampy carbon of statistical exposition into little memorable diamonds is an inspiration.

David Freedman, of the University of California at Berkeley, and his co-authors Robert Pisani and Roger Purves, are my sources for presenting the standard deviation as the "root mean square deviation" and the "likely size of chance error," for representing a distribution using a "box model," and for "standard units say how many SDs above or below the average a value is." Their textbook, *Statistics*, now in a second edition with Ani Adhikari as fourth author, remains after two decades perhaps the best-crafted textbook ever on the material it covers.

Frederick Mosteller, of Harvard University, who supervised my first teaching assistantship in graduate school, is the source of the shorthand that standard deviations "add like Pythagoras," and of the felicitous acronym BINS for remembering the four defining properties of the binomial distribution. Watching Fred teach was an education in itself.

Geoff Jowett (*Proceedings of the Fourth International Conference on the Teaching of Statistics*, pp. 15–23) first used the metaphor of an invisible man walking a dog to explain the difficult logic of confidence intervals. I like his metaphor so much that I have made a variant of it a unifying conceit for the Companion.

Finally, I would like to acknowledge the immense service to our profession provided by those authors who publish the original sources of the data sets they use in their books. By making it easier for others of us to short-cut the time-consuming search for effective examples from the real world, they have done much to raise the level of statistics teaching around the world.

Among the pioneers whose examples I have most often relied on here are:

Snedecor, George W. and William G. Cochran (1967). *Statistical Methods*, 6th ed. Ames, IA: The Iowa State University Press. (First edition 1937.)

Bliss, Chester I. (1970). *Statistics in Biology.* New York: McGraw Hill.

Larsen, Richard J. and Morris L. Marx (1986). *An Introduction to Mathematical Statistics and Its Applications*, 2nd ed. Englewood Cliffs, NJ: Prentice Hall. (First edition 1981.)

More recently, we also have an important collection of data sets:

Hand, D.J., Daly, A.D. Lund, K.J. McConway and E. Ostrowski (1994). *A Handbook of Small Data Sets.* London: Chapman and Hall.

an electronic companion to
statistics™

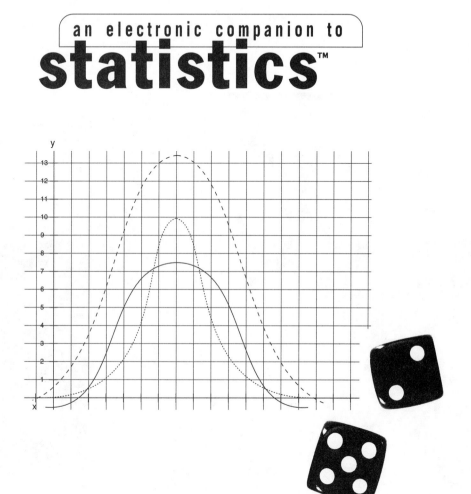

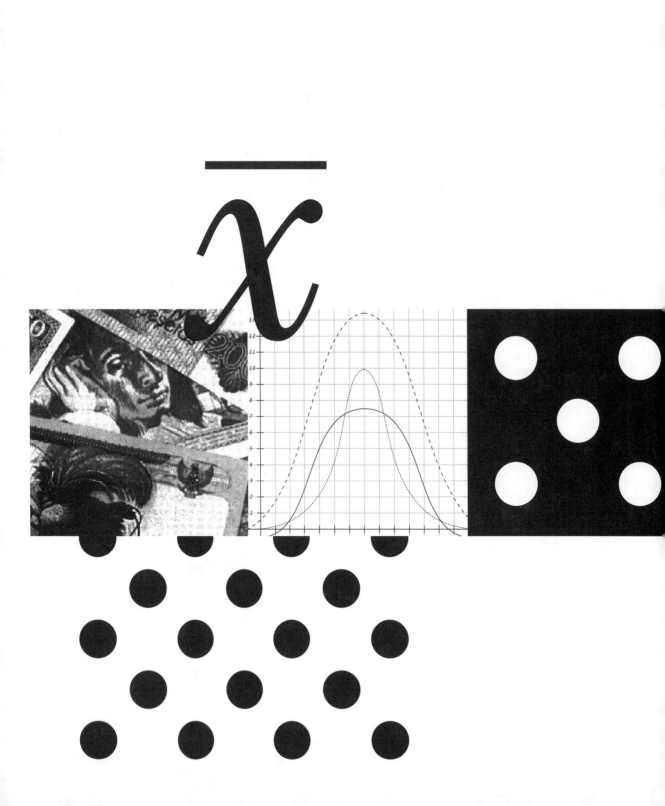

Data Basics

351.75	276.75	WaPost h	4.80	17	92	335.50	−2.13
19.88	17.75	WaWater	1.24	14	326	18.50	−.13
12.13	7.50	WastMln	..	..	1417	7.63	−.13
33.63	18.38	Waters	..	51	501	30.38	+.50
43.75	17.00	Watkln	.48	73	433	26.25	−.25
34.38	11.31	Watsco s	.14	33	116	29.50	−.63
24.63	15.50	Watts h	.31	..	363	24.38	+.25
6.63	1.00	Waxmn	..	3	786	5.88	−.13
38.13	23.13	WthfrdE	..	..	1012	35.63	−.13
20.00	15.25	WebbD	.20	..	274	16.50	..
36.50	23.50	Weeks n h	1.72	34	763	36.00	..
44.75	34.25	WeinRl	2.48	22	374	42.88	−.25
4.38	2.00	Weirt	..	..	901	3.00	..
34.88	28.75	WeisMk	.92	16	68	30.75	−.38
24.88	15.88	Wellmn	.32	20	666	17.63	−.25

Summary

1.1 IN A NUTSHELL

This is a short chapter, but the three sets of ideas it introduces form the foundation for the entire course and will come back again and again.

Cases and variables. **Cases** (or **individuals**) are the things that get measured: people, animal species, cities, colleges, years, etc. A **variable** is a feature of the cases: blood type (of people), life span (of animal species), bond rating (of cities), football conference (of colleges), inflation rate (of years). All the statistical methods you find in a typical first course rely on thinking about your data as a kind of spreadsheet, with cases as rows and variables as columns. (In fact, a major reason for the success of spreadsheet software like Excel and Lotus 1-2-3 is that the cases-by-variables structure is so natural for quantitative work.) However, data sets won't always come to you already in that form, so you need to work at recognizing cases and variables from the description of a data set.

Categorical versus quantitative. A **categorical** variable (blood type, football conference) tells which of several groups a case belongs to. A **quantitative** variable (life span, inflation rate) assigns a numerical value to each case. Why is this distinction important? Because it helps you decide what kind of statistical analysis to do. You can plot quantitative variables like life span and inflation rate on a number line; find the highest, lowest, and middle values; or compute the average value. You can't do any of those things with a categorical variable like blood type or football conference.

Distributions. Each variable, together with a collection of cases, gives a **distribution**. There is one value of the variable for each case, and the set of values, together with how often each value occurs, is the distribution. Distributions are important because they record the way a variable *varies* over a set of cases, and variability is what statistics is all about. If we didn't have variability, you wouldn't need statistical thinking. The entire second chapter is devoted to distributions—ways to explore for patterns and describe what you find—but in a sense, the whole of statistics is about distributions, so it is worth your effort to get to know them well.

1.2 HOW THESE IDEAS RELATE TO THE FOUR KEY THEMES

Production (and inference). The general question, "How and why were the data gathered?" can be made more precise by asking how cases and variables were chosen. Often the usefulness of data will depend critically on how the cases were obtained. Are they a complete set, as with the witches (Question 2)? The most extreme cases, as with the movies (Question 1)? Often we can't get a complete set, and then we need to choose a representative sample. (See Topic 6.) The way you choose the cases will determine whether or not *inference*—generalizing beyond the data you actually see—will be possible.

Exploration. Knowing the kind of variable can help guide a search for patterns, as you'll see in the next chapter.

Repetition. You can think of the distribution of a variable as the result of repetition: each case gives a value of the variable, and repeating (choosing additional cases) gives the distribution.

Self-Testing Questions

1. *Movies.* Identify cases and five quantitative variables in the box office report.

Top Weekend Movies
Weekend of Dec. 20–22, 1996
All dollar figures in millions
gross to date, weeks in
release, number of screens

	Weekend Gross
1. **Beavis and Butt-head Do America** $20.1, one week, 2,190 screens	$20.1
2. **Jerry Maguire** $36.3, two weeks, 2,531 screens	$13.1
3. **101 Dalmatians** $84.8, four weeks, 2,901 screens	$7.0
4. **Scream** $6.4, one week, 1,413 screens	$6.4
5. **One Fine Day** $6.2, one week, 1,946 screens	$6.2
6. **The Preacher's Wife** $15.3, two weeks, 1,989 screens	$5.2
7. **Mars Attacks!** $16.7, two weeks, 1,955 screens	$4.7
8. **Jingle All the Way** $47.4, five weeks, 2,112 screens	$3.1
9. **My Fellow Americans** $2.9, one week, 1,915 screens	$2.9
10. **Daylight** $20.4, three weeks, 1,990 screens	$2.4

SOURCE: Exhibitor Relations Co., Inc., *Daily Hampshire Gazette*, December 26, 1996.

Answer

Cases: movies

Variables: 1. weekend gross, in millions of dollars
2. gross to date, in millions of dollars
3. weeks in release
4. number of screens
5. rank, based on weekend gross

2. *Salem Witches.* Identify cases and variables. List the data in a rectangular table with cases as rows and variables as columns. Tell which variables are categorical and which are quantitative.

Nehemiah Abbot of Topsfield was accused of witchcraft on May 28, 1692. Daniel Andrew of Salem Village was accused on May 14, 1692. Bridget Bishop of Salem Village, accused on April 21, was tried and hanged. Elizabeth Carey, of Charlestown, was accused May 28. Giles Corey, of Salem Village, accused on April 18, was pressed to death. (A full listing would include 141 people, of whom 19 were hanged.)

Answer

| | VARIABLES | | | |
Cases	Date (Quantitative)	Place (Categorical)	Sex (Categorical)	Executed (Categorical)
Abbot, Nehemiah	5/28/92	Topsfield	M	*
Andrew, Daniel	5/14/92	Salem Village	M	*
Bishop, Bridget	4/21/92	Salem Village	F	Yes
Carey, Elizabeth	5/28/92	Charlestown	F	*
Corey, Giles	4/18/92	Salem Village	M	Yes

*not given (but in fact they were not executed)

Ambiguities about cases. For some data sets, there may be more than one way to think about cases.

Witches: grouped data. For the witch data of Problem 2, we could compute summaries by town. Each town would become a case. Possible variables would include:

Number of women accused
Number of women executed
Number of men accused
Number of men executed

A different summary might take the months (Jan. 1692, Feb. 1692, etc.) as cases.

3. *Diamond rings.* Identify cases and variables in the following advertisement for diamond rings. Which variables are categorical? Which are quantitative?

A.	1.25 Ct. Lil' Kohinoor	#1365 (Reg. $134) . . . . $ 85
	2.5 Ct. Lil' Kohinoor	#1366 (Reg. $177) $112
	3.0 Ct. Kohinoor	#3329 (Reg. $179) $125
	4.0 Ct. Kohinoor	#1360 (Reg. $313) $189
B.	2.0 Ct. Marquise	#3310 (Reg. $327) $199
	3.0 Ct. Marquise	#1392 (Reg. $360) $229
	3.0 Ct. in White Gold	#1392W (Reg. $380) . . $239
C.	1.0 Ct. Solitaire Ring	#1301 (Reg. $141) . . . . $ 89
	2.0 Ct. Solitaire Ring	#1303 (Reg. $164) $104
	3.0 Ct. Solitaire Ring	#1305 (Reg. $198) $119

SOURCE: ©1994 Singh Corporation, *USA Today*, August 16, 1995.

Answer

The cases are the various diamonds (one per row). The categorical variables are type (A,B, or C), name, and item number. The quantitative variables are weight (carets), regular price, and sale price.

4. The table below, from *The Wall Street Journal* (January 6, 1997), summarizes a week's worth of beer advertising on TV. Identify the cases and six variables. Tell whether each variable is categorical or quantitative.

Advertiser	Show (Network)	Date (Time)	% Viewers under 21
Coors Light	Hit List (BET)	Sept. 2 (8–10 p.m.)	51
Molson	Singled Out (MTV)	Sept. 2 (7 p.m.)	52
Molson Ice	Beavis and Butt-head (MTV)	Sept. 2 (11:30 p.m.)	48
Foster's	Singled Out (MTV)	Sept. 3 (11 p.m.)	46
Molson	Real World (MTV)	Sept. 3 (8:30 p.m.)	45
Foster's	Melrose Place (E!)	Sept. 2 (7–8 p.m.)	41
Miller	Unreal (BET)	Sept. 5 (8–10 p.m.)	65
Schlitz	Yo MTV (MTV)	Sept. 5 (10 p.m.)	50
Molson	Beavis and Butt-head (MTV)	Sept. 6 (10:30 p.m.)	69
Budweiser	Video Music Awards (MTV)	Sept. 7 (8:30 p.m.)	46

SOURCE: Competitive Media Reporting (ad placement), Nielson Media Research (viewer demographics).

Answer

The cases are the ten advertisers, one row for each case. There are three categorical variables: advertiser, show, and network. There are three quantitative variables: date, time and percentage of viewers under 21.

5. *Baseball attendance.* In 1995 many baseball fans were angry that players and owners took so long to agree on a contract. In a chart called "Slump at the Gate," *USA Today* listed the 28 major league baseball teams, together with average attendance figures for 1994 and 1995. Tell two reasonable ways to define cases and variables.

➤ *Solution*

If we take teams as cases, then our variables include 1994 average attendance and 1995 average attendance (and also percent change in attendance.) However, the chart in *USA Today* had two rows for each team: one row for 1994 and a second for 1995. Thus it is possible to think of each combination of team and year as a case, with average attendance as a variable. We would also want to include team and year as variables.

6. *Age discrimination.* Identify cases and variables. Tell which variables are categorical, which are quantitative.

> *In 1992, Robert Martin, who had been laid off by Westvaco Corporation, filed suit claiming age discrimination. Part of the data used in the case listed 36 salaried and 14 hourly employees. Here is one such listing: salaried employee number 33, a machine designer, was born in 1935 and hired in 1968. He was laid off in the second of five rounds of layoffs.*

Answer

Cases: the 50 employees
Variables: employee number
            job description
            year of birth
            year of hire
            when laid off

7. Make a distribution table based on the following data about the ages of adults who say they listen to big-league baseball on radio. Space the categories evenly with one of the categories being 55–64.

> *Ages of adults who listen to big-league baseball on the radio:*
> *18, 22, 26, 30, 31, 34, 35, 40, 41, 41, 43, 47, 48, 52, 58, 58, 67, 70.*

Answer

Ages of adults who say they
listen to big-league baseball
on the radio:

Age	Number	Percent
15–24	2	11%
25–34	4	22%
35–44	5	27%
45–54	3	16%
55–64	2	11%
65+	2	11%

8. A professor gave 2 exams and 5 quizzes to her class of 14 students and recorded the grades in her computer. What are the cases? What are the variables? How many variables are there?

Answer

The cases are the 14 students. The variables are the grades on the exams and quizzes, so there are seven variables.

9. A professor collected data from the students in his class. He recorded the following variables for each student: height, year in college, age, and whether or not the student has a job. Which of these are categorical variables?

Answer

Height and whether or not the student has a job are categorical variables.

10. The Lorain County administration office has death certificates for persons who have died in the county. These records list the name of the person, the age at death, and the cause of death. What are the cases? What are the variables?

Answer

The cases are the individual people who have died. The variables are the age at death and the cause of death.

11. Consider the data from Question 10. Which of the variable(s) are quantitative?

Answer Age at death is a quantitative variable.

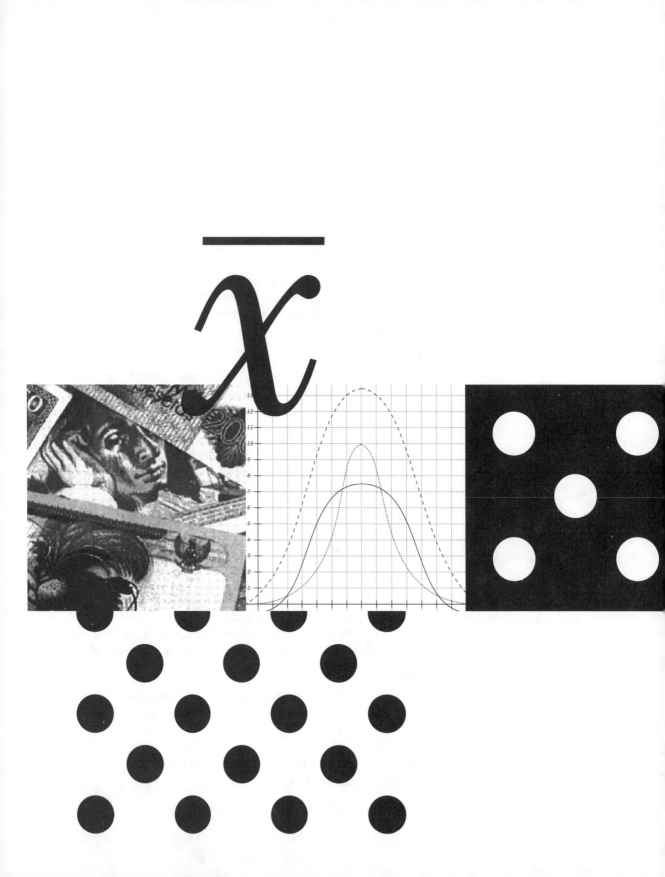

Describing Distributions

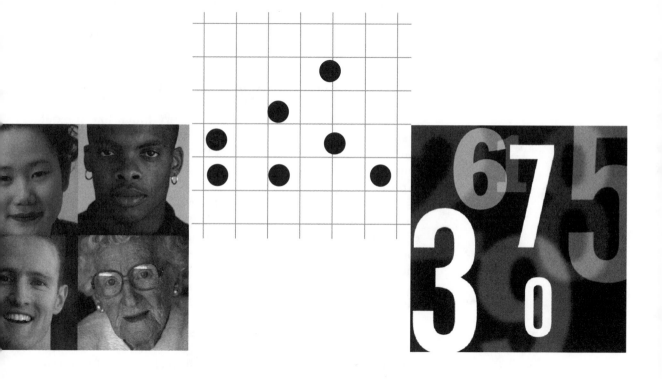

Summary _____

INTRODUCTION

Where does this topic fit in? This chapter introduces two sets of ideas that will be important throughout the whole of any beginning statistics course: distributions and exploring data.

Distributions. The idea of a distribution is perhaps the single most important idea in beginning statistics. If you pick a variable and look at its values for one case, then another, and another, and continue repeating this for a large number of cases, you find that the values **vary** from one case to the next. The **distribution** of the variable is just the collection of its values together with how often they occur. Because statistics is about learning in the presence of such variability, the concept of a distribution will come back again and again. The very next topic deals with the **normal distributions**, the single most important family of distributions in statistics. Topic 7 introduces **probability distributions**, which are very much like the data distributions of this chapter except they describe the chances that particular outcomes will occur. Probability distributions are important in statistics because the most reliable methods for data production (Topic 6) are probability based, and that fact allows us to describe the behavior of numerical summaries by giving their **sampling distributions** (Topic 8). These sampling distributions serve as the foundation for **inference** (Topics 9–13), the process you use to make valid generalizations that go beyond the limited amount of data you actually get to see. Either directly or indirectly, every chapter after this one relies on the idea of a distribution.

Exploration and description. This chapter sets a pattern that you'll find repeated in other chapters as well: **Plot — Shape — Summaries**. First, **Plot** your data to see its **Shape**. (If necessary, **Transform** to a new scale, and then re**Plot**.) Then choose suitable **Numerical Summaries**. For a distribution, the most important summaries are those that locate the **Center** of your distribution, and measure its **Spread**. If your distribution has a particular shape—symmetric and mound-shaped—the two summary numbers may tell you practically all you need to know about the distribution. In later chapters, which deal with relationships between two variables, your summaries will be different (for example, the slope of a fitted line, and a correlation coefficient), but the general pattern will be the same as before. Start with a **Plot**. Use the plot to think visually about your data and to categorize its **Shape**. Sometimes the shape will tell you that your data need to be **Transformed** to a new scale. But often, just as with distributions, the shape will be one that you can summarize with just a couple of numbers.

Note: Because the ideas and skills of this chapter are so fundamental, your success in your course depends on doing well with this material. To help make sure you get a solid foundation, we've made this chapter different from the others. Instead of a short summary and problems, we offer you a much more detailed explanation, with quick, simple exercises mixed in, so that you can be actively practicing your skills as you read along.

There are two main ways to summarize statistical data, numbers and pictures. Almost all pictures associate numbers with points on a line—a kind of glorified ruler. (If you're not already familiar with numbers as points on a line, check with your teacher for help. This is crucial!)

Checklist of Skills and Concepts

Introduction
    Checklist of Skills and Concepts
    Relation to Four Key Themes
Plots for Categorical Data
    Bar Charts
    Pie Charts
    shortcomings of pie charts
Plots and Shapes for Quantitative Data
describing shape: modes; skewness/symmetry; tails; gaps and outliers
    Dot Plots
    Stem Plots
    two-digit stems or leaves; rounding; splitting and pairing stems;
    parallel and back-to-back stem plots
Histograms
    y-axis: frequency, relative frequency; x-axis: interval width,
    starting value
Five-Number Summaries
    The Median
    The Quartiles
    Box Plots
    Modified Box Plots
    interquartile range; outliers
The Mean and Standard Deviation
    The Mean
    mean and total; deviations add to zero; mean as balance point;
    mean and symmetry
The Mean and Median Compared
    sensitivity to outliers
The Standard Deviation
    root mean square of the deviations
Properties of the SD
    SD = 0 means no variation; SD increases as spread increases;
    multiplying by a constant
SD and IQR Compared
    sensitivity to outliers

2.1 RELATION TO THE FOUR KEY THEMES

Production

How and why were the data produced? There are two connections here, although the first is more of a "non-connection"—there's nothing in a data distribution itself to tell you how the cases were chosen or how the variable was measured. The value of the data depends on these two things, which can't be known from the distribution alone. Second, looking ahead, the data distributions of this chapter are closely related to probability distributions (Topic 7), which tell the chances for various possible outcomes. As a rule, the best plans for producing data are those that can be described with probability distributions.

Exploration

Data = Pattern + Deviation There are two kinds of connections here. First, for comparing shapes of distributions, it can be useful to think of an ideal shape and deviations from the ideal. For most statistical work, the ideal is the bell (or normal) shape, which is the subject of the next chapter. You can think of other shapes as deviations from the ideal pattern: more than one mode, skewed, heavy-tailed, or with gaps and outliers. Second, for numerical summaries, you can think of the center (mean or median) as the pattern, and the spread (standard deviation or interquartile range) as a measure of deviation size.

Repetition

What will happen if I repeat this a large number of times? It can be useful to think of a distribution as the answer to this question. You choose a case and get the value of the variable for that case, then repeat the process a large number of times. The result is the distribution.

Interference

At this early stage, we're just exploring data, looking for patterns, and describing them. So far, there's no inference. However, three ideas from this chapter (tails, averages, and SDs) will be crucial later on. First, the tails of a mound-shaped distribution correspond to unusual, atypical outcomes. The basic idea of inference is that you should be suspicious of any theoretical distribution that would make your data be part of the tail. Second, the average and standard deviation, which in this chapter were defined for data distributions, will be useful later on for thinking about inference. If you have chosen a method for estimating some unknown number (the percentage who favor a political candidate, for example), it may be possible to describe the behavior of the estimation method in terms of its theoretical average and theoretical SD. A good method will have a small SD, and will have its theoretical average equal, or at least close to, the unknown number you want to estimate.

2.2 PLOTS FOR CATEGORICAL DATA

For categorical data, a distribution is typically given as a list of non-overlapping categories, with a number for each category telling either how often, or what percentage of the time, it occurred. There are two standard ways to show this information in a picture: a bar chart or a pie chart. Bar charts show numbers (how often, what percentage) using lengths, which make visual comparisons reasonably easy. Pie charts show percentages as angles, which are hard to judge visually. Almost any display that uses pie charts can be made more effective for conveying information by changing to bar charts.

Self-Testing Questions

1. Show the data A, A, B, B, B in a bar chart, using heights to show frequencies (counts).

Answer

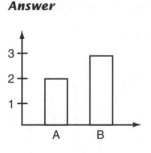

2. Show the same data (A, A, B, B, B) in a bar chart, this time using heights to show relative frequencies (percentages).

Answer There are five cases.

Category A: relative frequency = $2/5$ or 40%
Category B: relative frequency = $3/5$ or 60%

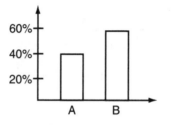

3. Show the data a, a, a, b, b, c in a bar chart, using heights to show frequencies.

Answer

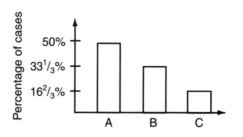

4. Show the same data (a, a, a, b, b, c) in a bar chart, this time using heights to show relative frequencies.

Answer

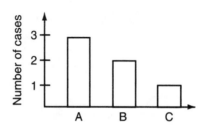

For showing a single distribution, putting the bars side by side makes visual comparisons easiest, but for comparing two or more distributions, we can put the bars end to end in a stacked bar chart.

5. Use stacked bar charts to compare Distribution 1: A, A, B, B, B and Distribution 2: A, A, A, B.

➤ *Solution*

First find relative frequencies.

Category	DISTRIBUTION 1			DISTRIBUTION 2		
	Count	Fraction	Percent	Count	Fraction	Percent
A	2	$2/5$	40	3	$3/4$	75
B	3	$3/5$	60	1	$1/4$	25

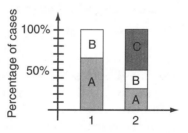

6. Use stacked bar charts to compare Distribution 1: A, A, B and Distribution 2: A, B, C, C.

Answer

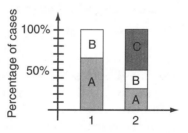

Although pie charts are not good at conveying numbers, they are used a lot anyway, and so it's useful to know how they are constructed.

7. Show the data A, A, B in a pie chart.

➤ *Solution*

First, find relative frequencies: A = $2/3$, B = $1/3$. Then divide a circle using the same fractions.

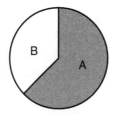

---

8. Show the data A, B, B, B in a pie chart.

Answer

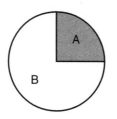

There are three general reasons why bar charts are better than pie charts.

1. If you have more than three or four categories, a pie chart becomes hopelessly messy, and you have to get all your information by "reading around the circle."
2. Sometimes several simple pie charts are used to show changes over time. Here, also, bar charts make such changes easier to see.
3. When the categories are defined numerically, pie charts are even worse than usual. You lose the natural order, and there are numbers all over the place.

2.3 PLOTS FOR QUANTITATIVE VARIABLES

For quantitative data, we'll consider four kinds of plots: dot plots, stem plots, histograms, and box plots. Box plots are different from the other three, and so we save them for a later section (2.4). The first three are in some ways quite similar. The dot plot is the simplest.

Once you've mastered the mechanics of the four kinds of plots, it's useful to think systematically about how they compare: what are the advantages and disadvantages of each plot. The table on p. 19 provides a comparison of that sort. You may also find it useful as an overview.

	Stem plot	Dot plot	Histogram	Box plot
How quick?	Quick	Moderate	Slower	Quick, once you have a stem plot
Size of data set?	Moderate to large	Small to moderate	Moderate to large	Moderate to large
Choice of formats?	Limited choices	One format works for all.	Many choices, some better than others.	One format works for all.
Versions, variations.	Regular, split or paired stems	Just one		Regular, modified
What it shows.	Each leaf is a case, width of set of leaves shows count.	Each dot is a case, density of dots show frequency.	Areas tell percent of cases; if bars are of equal width, height tells number of cases.	The five-number summary: min, Q_1, median, Q_3, max
Does it give up information?	No	Only through rounding off	Some	Yes; modes and clusters don't show.
Uses?	A quick, almost-histogram; shows shape and makes it easy to find summaries based on rank order	Comparing several groups	A good format can show shape well but histograms are mainly a device to help you think about distributions	Comparing several groups

Dot Plots

9. Construct a dot plot for this set of numbers: 0, 5, 5, 10, 20.

Answer

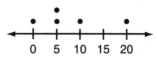

There are only two kinds of decisions when drawing a dot plot. First you need to decide where to start, where to end, and how many tick marks to use. Just use common sense, and be willing to redo a plot if you don't like the way it looks.

10. Construct a dot plot for the data 318, 301, 241, 18, 117.

Answer

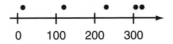

11. Construct a dot plot for the data 318, 301, 310, 315, 315.

Answer

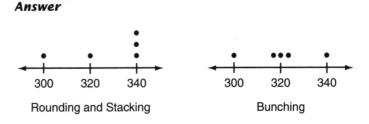

The second kind of decision is how to show crowding—either by rounding and stacking or by bunching. Here, too, just use your common sense. (Rules would make it too complicated.)

12. Construct at dot plot for the set of numbers 318, 320, 319, 300, 340.

Answer

If you have two or more groups of cases, you can use a parallel dot plot to compare groups.

13. Construct parallel dot plots for these two sets of numbers.

A: 0, 5, 5, 10
B: −5, 0, 0, 10

Answer

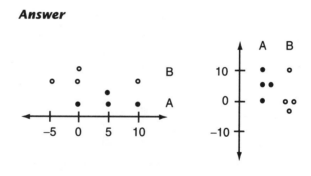

14. Construct a parallel dot plot for these two sets of numbers.

A: 10, 80, 100, 100
B: 40, 50, 60, 70

Answer

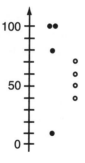

15. *Baseball Standings.* The data show the percentage of games won in 1995 by the teams of the three divisions of the American League.

Eastern		Central		Western	
Boston	59.7	Cleveland	69.4	Seattle	54.5
New York	54.9	Kansas City	48.6	California	53.8
Baltimore	49.3	Chicago	47.2	Texas	51.4
Detroit	41.7	Milwaukee	45.1	Oakland	46.5
Toronto	38.9	Minnesota	38.9		

(a) Construct a parallel dot plot for the data.
(b) On balance, which division was strongest?
(c) Which division was most evenly balanced?
(d) One division had an obvious winner long before the season ended. Which division was it?

Answer

(a)

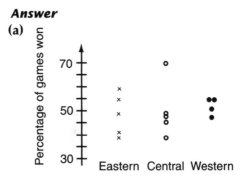

(b) The answer depends on what you mean by strongest, but the average percentage is highest for the Western division.

(c) Western

(d) Central

16. *Osmoregulation.* Some species of sea worms are able to maintain a higher salt concentration in their bodies than in the surrounding water. Other species ("non-regulators") don't have this ability. If placed in water with low salt concentration, they absorb additional water until their internal concentration matches that of their environment. Here are weights (as a percent of original weight) for 12 sea worms after 90 minutes in 33% sea water, 67% fresh water.

Nereis virens: 152, 138, 156, 129, 129, 155
Goldfingia gouldii: 188, 158, 179, 176, 132, 191

(a) Draw a parallel dot plot for the two species.
(b) On average, which species shows the greater percentage increase in body weight?
(c) On average, which species shows the greater variability?
(d) Which species is the "non-regulator" and why?

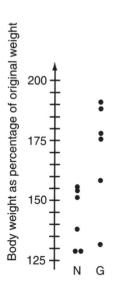

Body weight as percentage of original weight

200

175

150

125

N G

Answer

(a) See parallel dot plot at left.
(b) Goldfingia
(c) Goldfingia
(d) The non-regulator is Goldfingia, the species whose body weight is more affected by the change in environment.

17. *"Uh...Science, Anyone?"* Psychologists at Columbia University recorded several lectures in each of ten subjects, and counted the number of times per minute the lecturer said "uh" or "ah" or "um" or another pause filler. The researchers theorized that speakers fill pauses when they are searching for the next word, and that in subjects where there are precisely worded definitions, speakers will search less often and so say "uh" less often.

Natural Sciences: .97 (biology), 1.62 (chemistry), 1.30 (math), 1.8 (psychology)
Social Sciences: 2.54 (economics), 5.61 (political science), 3.73 (sociology)
Humanities: 6.06 (art history), 6.54 (English literature), 1.65 (philosophy)

(a) Draw a parallel dot plot.
(b) Compare the centers of the distributions for the three groups.
(c) Compare the spreads of the three distributions for the three groups.
(d) Does the pattern of differences between the three groups of subjects tend to support the researchers' theory?
(e) Does the theory provide a possible explanation for the low value in the social sciences and the lowest value in the humanities?

Answer

(a)

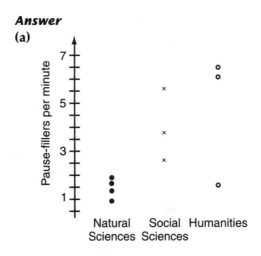

(b) Natural Science is lowest, Social Science is next, the Humanities is highest.

(c) Spreads follow the same pattern as centers: smallest for the Natural Sciences, largest for the Humanities.

(d) Yes.

(e) Yes. Economics is the most quantitative of the Social Sciences listed, and Philosophy is the subject in the Humanities for which precision about definitions and the use of words is most like the Natural Sciences.

18. *Stock Markets in Europe.* Recent years have seen U.S. investors become more interested in overseas stock markets. The data here show percentage change in dollar value, over a single month, for markets of Europe and the Mediterranean.

Northern Europe		Southern Europe and Mediterranean	
Belgium	4.7	France	3.3
Denmark	.1	Italy	2.6
Finland	2.2	Israel	−5.4
Germany	2.7	Greece	−.7
Ireland	−.1	Jordan	−3.1
Netherlands	.5	Portugal	−10.4
Norway	1.2	Spain	−2.3
Sweden	2.1	Turkey	−25.6
United Kingdom	.1		

(a) Construct a parallel dot plot.

(b) How do the distributions compare?

Answer

(a) Percentage change in dollar value −25 | 6 = −25.6%

Northern Europe			Southern Europe
	−25	6	
	−24		
	−11		
	−10	4	
	−9		
	−8		
	−7		
	−6		
	−5	4	
	−4		
	−3	1	
	−2	3	
	−1		
1	−0	7	
5 1 1	0		
2	1		
7 2 1	2	6	
	3	3	
7	4		

(b) The values for Northern Europe are almost all positive, and bunch together between 0 and 5%. The values for Southern Europe are much more spread out, and three-fourths are negative.

Stem Plots

Dot plots have a few disadvantages. It's hard to go from the dots back to the numbers, the plot can get crowded if you have more than a dozen or two cases, and the plots don't show the degree of bunching quantitatively. Stem plots overcome all three disadvantages. Cases are written as numbers, and bunching gets turned into "sideways stacking," which avoids clutter when you have many cases, and also lets you use width to judge the degree of bunching. (To get these advantages, you have to learn several variations on the basic plot, and learn to choose suitable variations.)

19. Construct a stem plot for the numbers 42, 19, 52, 11, 37, 15. (Write 42 as 4 | 2.)

➤ *Solution*

	1	159	(Bunching shows as 3 leaves
	2		on the same stem.)
	3	7	
(4 = stem)	4	2	(leaf = 2)
	5	2	

20. Construct a stem plot for the numbers 31, 28, 24, 35, 18. (Write 31 as 3 | 1.)

Answer

```
1 | 8
2 | 48
3 | 15
```

Sometimes you use two-digit stems.

21. Construct a stem plot for the numbers 318, 329, 302, 336, 309. (Write 318 as 31 | 8.)

Answer

```
30 | 29
31 | 8
32 | 9
33 | 6
```

If the important variation is mainly in the next-to-last digit, you may want to round off.

22. Construct a stem plot for the numbers 318, 329, 406, 519, 602. (Round 318 to 320 and write as 3 | 2.)

Answer

```
3 | 23
4 | 1
5 | 2
6 | 0
```

An alternative is to use two-digit leaves.

23. Construct a stem plot for the numbers 614, 673, 591, 308, 416. (Use two-digit leaves and write 614 as 6 | 14.)

Answer

```
3 | 08
4 | 16
5 | 91
6 | 14 73
```

You can spread out a stem plot by using split stems. Instead of ten kinds of leaves per stem, you have five.

24. 89, 84, 86, 91, 97, 85, 75, 76 (Use split stems.)

➤ **Solution**

```
7*  |  56          •stem for leaves 0–4
8•  |  4           * stem for leaves 5–9
8*  |  569
9•  |  1
9*  |  7
```

25. Construct a stem plot for the numbers 61, 63, 65, 58, 69. (Use split stems.)

Answer

```
5•  |
5*  |  8
6•  |  13
6*  |  59
```

Another variation uses paired stems, with two kinds of leaves per stem.

26. Construct a stem plot for the numbers 79, 84, 86, 91, 87, 85, 78, 82. (Use paired stems.)

➤ **Solution**

```
7*  |  89          • for leaves 0, 1
8•  |              t for leaves two, three
 t  |  2           f for leaves four, five
 f  |  45          s for leaves six, seven
 s  |  67          * for leaves eight, nine
8*  |
9•  |
```

27. Construct a stem plot for the numbers 61, 63, 65, 58, 69. (Use paired stems.)

Answer

```
5* | 8
6• | 1
 t | 3
 f | 5
 s |
6* | 9
```

28. Rewrite the plot below by splitting stems to stretch it.

```
3 | 0488
4 | 17
5 | 0089
```

Answer

```
3• | 04
3* | 88
4• | 1
4* | 7
5• | 00
5* | 89
```

29. Rewrite the plot below by splitting stems to stretch it.

```
5 | 13367
6 | 0488
7 | 29
```

Answer

```
5• | 133
5* | 67
6• | 04
6* | 88
7• | 2
7* | 9
```

30. Rewrite the plot below by pairing stems to make it more compact.

```
 2 | 0
 3 | 1
 4 |
 5 | 09
 6 | 1
 7 | 3
 8 | 1
 9 |
10 | 3
```

Answer

```
0• |
 t | 20 31
 f | 50 59
 s | 61 73
0* | 81
1• | 03
```

31. Rewrite the plot below by pairing stems.

```
10 | 8
11 |
12 | 345
13 |
14 | 6
15 | 7
16 | 0
```

Answer

```
1• | 08
 t | 23 24 25
 f | 46 57
 s | 60
1* |
```

When you have two or more groups of cases, you can construct parallel stem plots to compare the groups visually.

32. Draw a parallel stem plot.

Group A: 5, 9, 18, 20
Group B: 15, 25, 31, 37

Answer

Group A		Group B	
0	59	0	
1	8	1	5
2	0	2	5
3		3	17

If you only have two groups, you can share one set of stems in a back-to-back stem plot.

33. Rewrite the parallel stem plots back-to-back.

Answer

Group A   Group B

95	0	
8	1	5
0	2	5
	3	17

34. Draw parallel stem plots to compare Groups A–C.

Group A: 0, 1, 6, 12
Group B: 6, 12, 19, 20
Group C: 21, 30, 35, 40, 48

Answer

A		B		C	
0	016	0	6	0	
1	2	1	29	1	
2		2	0	2	1
3		3		3	05
4		4		4	08

35. Rewrite the stem plots for Groups A and C back-to-back.

Answer

```
   A     C
610 | 0 |
  2 | 1 |
    | 2 | 1
    | 3 | 05
    | 4 | 08
```

36. *How Expensive is New York?* Here are average costs of hotel rooms in a dozen major U.S. cities in June 1995. New York costs the most, and a stem plot helps compare its cost with other cities' costs.

Average Cost of a Hotel Room in a Dozen Big-City Destinations Nationwide

Atlanta	$ 69.21
Boston	$ 99.56
Chicago	$ 86.72
Denver	$ 67.05
Honolulu	$ 101.18
Los Angeles	$ 70.53
Miami	$ 68.65
New Orleans	$ 80.33
New York	$ 129.19
Orlando	$ 65.29
San Francisco	$ 94.11
Washington, D.C.	$ 90.17

(a) Round off the costs to the nearest whole dollar, and show the distribution in a stem plot. Use tens of dollars as stems so that, for example, $69 = 6 | 9.

(b) Show the same data in a dot plot. Notice how the dense cluster of dots around $70 corresponds to the widest row (6 | 5 7 9 9) of your stem plot.

(c) There is a second cluster that is more spread out than the one near $70. How does it show up in the stem plot?

(d) How far is it from the highest cost (New York) to the next highest?

Answer

(a)
```
  6 | 5799
  7 | 1
  8 | 07
  9 | 04
 10 | 01
 11 |
 12 | 9
```

(b)

Cost of hotel room

(c) The second cluster shows up as a set of three stems in a row with two leaves each.

(d) About $28 ($129–$101)

---

37. *How Bad is TWA?* USAir, Southwest, and Alaska Airlines come out at the top of two lists that measure airline performance; TWA rolls up to the gate behind all the others.

Flights That Arrived On Time		Complaints per 100,000 Customers	
Alaska	89.0%	Southwest	.19
USAir	84.9%	Alaska	.51
Southwest	83.6%	USAir	.61
Northwest	83.4%	Northwest	.62
United	81.5%	American	.74
America West	80.8%	Delta	.85
Delta	79.0%	United	.86
Continental	78.2%	Continental	1.16
American	76.8%	America West	1.18
TWA	68.1%	TWA	1.31

(a) Show the complaints per 100,000 customers in a stem plot. (Write 0.19 as 1 | 9.)

(b) Based on your plot, how many airlines would you judge to be unusually "bad" (unusually likely to generate complaints)? How many would you judge to be unusually "good" (few complaints)?

(c) Notice that the choice of stems in part (a) makes for a long, straggly plot. Redo your stem plot using paired stems and two-digit leaves. Which plot do you prefer?

(d) Now, construct a stem plot for on-time percentages. Round to the nearest whole percentage point and use 6, 7, and 8 as stems. (Write 89% as 8 | 9.)

(e) The plot in part (d) is too bunched together to show the distribution well. Redo the plot using split stems 6∗, 7•, 7∗, 8•, and 8∗.

(f) According to on-time percentages, how many airlines would you judge unusually "good"? Unusually "bad"?

(g) Why can't you use a back-to-back stem plot here?

(h) The analysis based on stem plots alone can't use some of the important information in the data. What do you consider the most important unused information?

Answer

(a)

```
              1 | 9
              2 |
              3 |
              4 |
    s         5 | 1
  r           6 | 12
  e           7 | 4
  m           8 | 56
  o           9 |
  t          10 |
  s          11 | 68
  u          12 |
  c          13 | 1          1|9 = 0.19
Complaints per
100,000 customers
```

(b) Either one airline, or three, might be judged unusually bad, and one airline, Southwest, is unusually good.

(c)

```
              0• | 19
               t |
               f | 51
               s | 61 62 74
              0* | 85 86
              1• | 16 18
               t | 31              0•|19 = 0.19
Complaints per
100,000 customers
```

(d)

```
On-time
percentage   6 | 8
             7 | 789
             8 | 123459    6|8 = 68%
```

(e)

```
             6• |
             6* | 8
             7• |
             7* | 789
             8• | 1234
             8* | 59        6*|8 = 68%
On-time
percentage
```

(f) One airline (Alaska at 89%) might be judged unusually good. One airline (TWA at 68%) was unusually bad.

(g) There are two variables. You can use a back-to-back stem plot if you have one variable and two groups of cases.

(h) The data come in pairs, with two variables reported for each airline (case).

38. *Mantle's Averages.* Mickey Mantle played 18 regular seasons for the New York Yankees and was in the World Series 12 times. During a regular season he would have about 450 official times at bat. During a World Series he averaged about 19 times at bat. The data in the table show his batting averages (fraction of hits).

(a) Before looking at the data, guess which set of averages, regular season or World Series, will tend to be higher. Then guess which set will tend to be more spread out.

Year	Regular Season	World Series
1951	.267	.200
1952	.311	.345
1953	.295	.208
1954	.300	
1955	.306	.200
1956	.353	.250
1957	.365	.263
1958	.304	.250
1959	.285	
1960	.275	.400
1961	.317	.167
1962	.321	.120
1963	.314	.133
1964	.303	.333
1965	.255	
1966	.282	
1967	.245	
1968	.237	

(b) Construct a back-to-back stem plot to compare the two sets of averages. Use split stems and two-digit leaves.

(c) Which set tends to be higher? (Is this what you expected?)

(d) Which set is more spread out? Can you guess the main reason?

(e) A comparison based on the back-to-back stem plot ignores (or at least doesn't use) important information in the data. What information is that?

Answer

(a) *Averages.* You might guess that the motivation of the World Series would make series averages higher than regular season averages. On the other hand, the limited schedule—at most seven games—allows the opposition to use their best pitchers more often, which would lead to lower series averages. *Spread.* Regular season averages are based on much larger sample sizes (times at bat). The larger **n** is, the smaller the SD of an average tends to be.

(b)

World Series		Regular Season
33, 20	1•	
67	1∗	
08, 00, 00	2•	37, 45
63, 50, 50	2∗	55, 67, 75, 82, 85, 95
45, 33	3•	00, 03, 04, 06, 11, 14, 17, 21
	3∗	53, 65
00	4•	

(c) Regular season averages tend to be higher.

(d) World Series averages are more spread out because they are based on smaller samples (fewer at bats).

(e) The data come in pairs, according to year.

39. *Salty Chips.* Most people get a lot more sodium in their diet than they need, but many of us like the taste of salt, especially on something like tortilla chips. Is salt necessary for good taste? *Consumer Reports* published a comparison of brands of tortilla chips according to quality (flavor and texture) and sodium content. Here are sodium levels (mg/oz.) for 29 brands, grouped according to quality.

Excellent: 55, 80, 80, 70, 75, 125, 38, 170, 50, 65, 85, 75, 80, 135, 132, 140
Good: 85, 0, 198, 110, 160, 5, 55, 99, 76, 26, 130, 170

(a) Show the sodium levels in a back-to-back stem plot. (What stems to use is not obvious. You may find you don't like the choice you start with. Don't hesitate to revise.)

(b) Based on your stem plot, would you say there's a clear difference in sodium levels between the chips rated "good" and those rated "excellent"?

Answer

(a)   Salt Content of Chips   0*|80 = 80 mg./oz

Good		Excellent
05, 00	0•	
26	t	38
55	f	50, 55
76	s	65, 70, 75, 75
99, 85	0*	80, 80, 80, 85, 85
10	1•	
30	t	25, 32, 35
	f	40
70, 60	s	70
98	1*	

(b) No

40.   *Where the Jobs Are.* The AAUP data set (Data Appendix 1, p. 78) shows percent unemployed in each of 28 areas of graduate or professional study.
   (a) Construct a stem plot of the unemployment rates. Use whole percents (0, 1, . . . , 5) as stems (so, for example, 0.7% is written 0 | 7).
   (b) Describe the shape: How many modes? Symmetric or skewed? What kind of tails? Are there any outliers?

Answer

(a) Percent unemployed 0 | 1 = 0.1%

0	1233455789
1	111124458999
2	11
3	15
4	3
5	5

(b) The distribution of unemployment rates for the 28 academic fields has one mode, and is skewed toward high values, with a long right tail, but no outliers. The highest rates of unemployment are in drama, music, and foreign languages.

The choice of stems can sometimes make a big difference in what features of a distribution stand out in a stem plot. Here's such an instance.

41. *Where the Dollars Are.* Which subjects pay best? Use the average annual salaries for faculty jobs from the AAUP data set.
   (a) Construct a stem plot. Round each salary to the nearest thousand and use split stems (2•, 2∗, 3•, 3∗, 4•). Think how you would describe the resulting shape.
   (b) Now expand your stem plot using paired stems (2f, 2s, 2∗, 3•, 3t, 3f, . . . , etc.). What features stand out more clearly in the second plot?

Answer
(a) Average annual salary 2• | 4 = $24,000

2•	4
2∗	55666778999
3•	00011123333
3∗	67
4•	134

(b) Average annual salary

2•	
t	
f	455
s	66677
2∗	8999
3•	000111
t	23333
f	
s	67
3∗	
4•	1
t	34

Straggling large values are much easier to see in this plot. The top three salaries are for dentistry, medicine, and law; the next two are for agriculture and engineering.

42. *Where the "Women's Jobs" Are.* Now refer to Question 41 and work with the percentage of women in faculty jobs in the same 28 fields.
   (a) Construct a stem plot. (Round to the nearest whole percent, and write 16% as 1 | 6.)
   (b) What do you consider the most striking aspects of the shape of your plot?

(c) Now, expand the lower part of your plot for percentages below 60%, using split stems (0*, 1•, 1*, 2•, ..., etc.).

(d) What features stand out more clearly in your second plot?

Answer

(a) Percentage of women in faculty jobs, by field 0 | 5 = 5%

```
0 | 57
1 | 345566
2 | 367
3 | 0124
4 | 1668
5 | 024588
6 |
7 |
8 | 02
9 | 4
```

(b) The feature that stands out most is the set of three fields for which most faculty are women: library science, nursing, and social work.

(c) Percentage by field

```
0• |
0* | 57
1• | 34
1* | 5566
2• | 3
2* | 67
3• | 0124
3* | 1
4• | 668
4* | 024
5• |
5* | 588
```

(d) The distribution is "lumpy" with several small clusters. If you go back to the data, you can see that the clusters have a logic to them. For example, the lowest cluster corresponds to the most quantitative subjects.

Sometimes a distribution with a striking shape can lead to useful ideas about possible explanations.

43. *Where the Non-Academic Jobs Are.* Refer again to Question 41 and look at the percentages of non-academic jobs in each subject.

   (a) Construct a stem plot. Round to the nearest whole percent and use 0, 1, . . . , 9 as stems.
   (b) Describe the shape of the plot. (Did you guess what the shape would be before you made the plot?)
   (c) What does the shape of the plot suggest to you about (meaningful groupings within) the 28 subjects?

Answer

(a) Percentage of jobs that are non-academic. 1 | 2 = 12%

```
1 | 22378
2 | 17
3 | 4
4 | 13
5 | 18
6 | 26
7 | 9
8 | 3
9 | 366777899999
```

(b) The distribution is U-shaped, with modes at both extremes and few values in between.

(c) In some fields, most jobs are academic; for a dozen fields, almost all jobs are non-academic.

With any statistical analysis, it is always worth asking whether you have taken advantage of all the available information.

44. *Where Additional Information Is.* Answers to the last few problems, taken together, tell a lot about the 28 cases and 4 variables you have been plotting. However, these analyses don't take advantage of one very important aspect of the structure of the data. What is this feature, and why is it important?

Answer

These plots show only one variable at a time. They tell us nothing about the relationships between pairs of variables. So they can't help answer such questions as, "Are salaries higher in fields with lower unemployment rates?" "Are faculty salaries higher in fields that employ fewer women?"

45. *Old Airplanes.* For some commercial airlines, the average age of the fleet makes the planes older than most college freshmen. Use the ages listed below to compare B-727 airplanes with B-737s via a back-to-back stem plot. Is there a clear difference in average ages?

Model	Airline	Average Age of Fleet	Number of Planes
B-727	American	18.2	80
	Continental	18.6	48
	Delta	18.2	134
	Kiwi	22.1	16
	Northwest	16.6	52
	TWA	21.3	41
	United	16.3	74
	USAir	17.7	6
	USAir Shuttle	25.0	10
B-737	Air South	20.8	8
	Airtran	19.3	4
	Alaska	5.4	30
	Aloha	12.0	17
	America West	10.7	56
	Continental	10.9	123
	Delta	10.3	67
	Markair	7.2	7
	Nations Alr Express	26.6	2
	Southwest	7.8	210
	United	10.5	227
	USAir	9.2	216
	Vanguard	25.8	5
	Western Pacific	9.0	5

Answer

Average Age of Fleet   1*|6 = 16 years

```
        B737          B727
              | 0• |
     9 9 8 7 5| 0* |
     2 1 1 1 0| 1• |
             9| 1* | 6 7 8 8 8 9
             1| 2• | 1 2
           7 6| 2* | 5
```

Overall, the 727s tend to be older.

46. *Radioactive Twins.* Years ago, scientists who wanted to know if city living was bad for human lungs carried out an unusual study. Incredibly, they managed to find seven twin pairs with one in each pair living in the country, one in the city.

Even more incredibly, they got both twins in each pair to inhale an aerosol of radioactive Teflon particles! This allowed them to measure the percentage of particles still in the lungs an hour later. Here are the percentages.

Twin Pair	I	II	III	IV	V	VI	VII
Rural	10.1	51.8	33.5	32.8	69.0	38.8	54.6
Urban	28.1	36.2	40.7	38.8	71.0	47.0	57.0

(a) Show the data in a back-to-back stem plot.
(b) Evaluate the pattern: Would you say that the rural twins tend to clear their lungs faster (lower percentage remaining) than the urban twins?
(c) This analysis leaves out important information. What do you consider the most important omitted information? Why is it important?

Answer

(a) Percentage of
Radioactivity
After One Hour   1|0 = 10%

```
    Urban         Rural
           |  1 | 0
         8 |  2 |
       9 6 |  3 | 3 4 9
       7 1 |  4 |
         7 |  5 | 2 5
           |  6 | 9
         1 |  7 |
```

(b) Although the data suggest that percentages tend to be lower for those who live in the country, the pattern is not striking or persuasive.
(c) The omitted information is the pairing—the numbers on the left and right side are grouped into twin pairs, but the plot doesn't show this.

Histograms

A histogram is similar to a stacked dot plot, or a stem plot rotated 90°.

Data: 21, 22, 23, 30

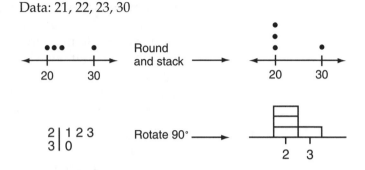

The horizontal axis uses a number line to show values of a quantitative variable. The vertical axis shows how often the values occur.

Histograms serve two main purposes. First, they can sometimes show the shape of a distribution better than a stem plot can. Second, learning about histograms for data can make it easier to learn about probability histograms later on.

47. Draw a histogram to represent the following list of numbers. Use intervals 0–9, 10–19, etc., and show frequency (number of cases) on the *y*-axis.

Data: 7, 8, 10, 11, 11, 14, 18, 31, 45, 46

➤ *Solution*

Step 1. *Count* the number of cases in each interval.

Interval	Count
0–9	‖
10–19	₩
20–29	
30–39	│
40–49	‖

Step 2. *Draw* a rectangle over each interval on the number line, with height = count.

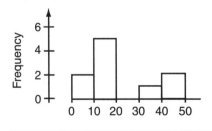

48. Construct a frequency histogram for the data that follow. Use intervals 0–4, 5–9, 10–14, 15–19, 20–24, and 25–29.

Data: 0, 1, 6, 6, 7, 10, 13, 15, 16, 28

Answer

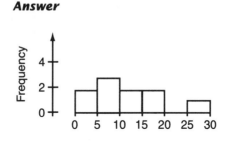

49. *Accused Witches 1.* The Salem Village witchcraft hysteria of 1692 eventually saw 19 accused witches executed. The data suggest, however, that the first of the executions had a noticeable effect on the pace of the accusations. (Before looking at the data, try to guess the likely effect of a public hanging in a small community.)

Here are the accusations, by month.

Month	Accusations
February	3
March	5
April	23
May	37
June	6
July	8
August	20
September	20
October	1
November	3
?	13

(a) Draw a time plot. The first hanging (of Bridget Bishop) took place on June 10, 1692. What does the pattern in the plot suggest about the effect of the execution on the accusations?

(b) Construct a frequency histogram.

(c) Can you guess about when the first execution took place?

Answer
(a), (b)

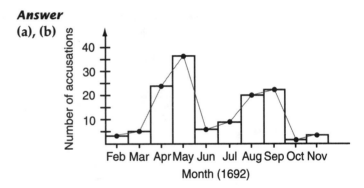

(c) The first execution was in June. After Bridget Bishop went to the gallows, accusations were less frequent, for a short time.

A relative frequency histogram is like a frequency histogram except that it shows fraction or percentage of the total count on the *y*-axis.

50. Construct a relative frequency histogram for the data 0, 1, 2, 7. Use intervals 0–3, 4–6, 7–9.

➤ **Solution**

There are four cases in all—three (75%) in the first interval and one (25%) in the third interval.

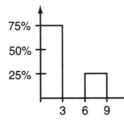

51. *Accused Witches 2.* In fairy tales, the typical witch is an older woman. Is the fictional stereotype a mirror of historical reality in Puritan New England? Between 1620 and 1725, accusations of witchcraft were leveled at 156 women whose ages could be determined. Here is the distribution of ages.

Age	<10	10–19	20–29	30–39	40–49	50–59	60–69	70–79	80+
Number Accused	3	23	26	23	31	23	19	7	1
Percent	1	15	17	15	20	15	12	4	.6

(a) Use the percentage data to construct a relative frequency histogram. (Assume, for convenience, that 80+ means 80–89.)

(b) How good do you consider the fit between data and stereotype?

Answer

(a)

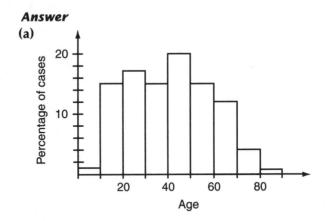

(b) If the ages fit the stereotype, there would be fewer cases on the left, more cases on the right. In fact, the age distribution is fairly uniform.

52. *Hiring Patterns 1: Months.* When during the year (what months) would you expect a company to do most of its hiring? (Think about factors affecting supply and demand for workers, and try to guess.) Here is a summary table for the month of hire for 50 employees in one department at the Envelope Division of Westvaco Corporation.

Month	Jan	Feb	March	April	May	June	July	Aug	Sept	Oct	Nov	Dec
No. Hired	1	8	6	4	3	3	4	6	7	2	3	3

(a) Construct a relative frequency histogram.

(b) Describe the pattern, paying particular attention to peaks (modes) and valleys. Which months see the most new hires?

(c) What percentage of new hires took place in the two months with the most new hires?

(d) Think up a believable explanation. (Your explanation might be right or wrong. As long as it accounts for the pattern, there's no way to know just from the data whether you're right.)

(e) The data in this problem may not be representative of employees hired at other companies or, for that matter, in other departments at Westvaco. In fact, there are ways in which these numbers may not even be representative of hiring patterns in the department where these employees worked. Can you guess what might make them a biased sample?

Answer

(a) There are 50 employees, so each case corresponds to 2% of the total.

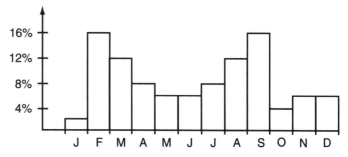

(b) The distribution is bimodal, with peaks in Feb–Mar and Aug–Sep.

(c) 31%, or almost one-third of new hires took place in February or September.

(d) One possibility is that hiring is tied to the academic calendar. Another is that hiring is tied to the company's financial planning. A third is that the modes occur at the ends of traditional vacation times (after Dec–Jan and Jul–Aug).

(e) There are many factors that may make these employees atypical, and so biased as a sample of all employees. Here's one: These are employees who don't represent the ones who have already left the company for one reason or another.

Choosing intervals. Because you can choose how many intervals to use, how wide to make them, and where to start the first one, histograms are more flexible than stem plots. Unless you have a lot of data, however, different choices for intervals can lead to histograms with somewhat different shapes.

53. *Hiring Patterns 2: Years.* Do you imagine that most of the 50 employees from Question 52 were hired recently? A long time ago? Somewhere in between? The actual pattern for the Westvaco data is more complicated than you might guess. Use the Martin data set (Data Appendix 2, p. 79) for this problem.

(Note: This problem is deliberately designed to use your time inefficiently in order to drive home an important point. By all means, read the whole problem before you start work so you can figure out how to do things more efficiently.)

(a) Construct a histogram for the variable hire year using intervals 40–49, 50–59, etc.

(b) Now do another histogram using narrower intervals 40–44, 45–49, 50–54, etc.

Notice that you have to start over from scratch. There are two lessons here. First, histograms are a lot slower than stem plots, and are best left to a computer. Second, if you have to draw histograms by hand, start with a stem plot; it will be much easier to redraw your histogram if you want to use different intervals.

(c) The second plot, with narrower intervals, shows a clear bimodal pattern. Guess at a believable explanation based on a change in company hiring policy.

(d) Guess at a different explanation based on assuming that there are two groups of employees: one group who feel no commitment to the company, and another group "in for the long haul."

(e) Can you think of a way to use the data to decide whether part (c) or part (d) is more nearly correct?

Answer

(a) Histograms are easier to draw if you first make a stem plot:
Year of hire 4 | 3 = 1943

```
4 | 3
5 | 1234559
6 | 0222334445667778
7 | 12334467788
8 | 13555566789999
9 | 0
```

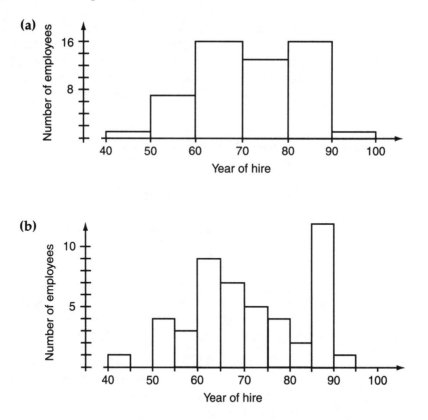

(c) One way to get a pattern like this would be for the company to hire many employees in the 1960s and to hire many more in the late 1980s.

(d) An alternative explanation: The mode in the 1960s corresponds to employees who are in it for the long haul. The spike in the late 1980s corresponds to employees who work for a few years, and then move on.

(e) The data in the histograms can't tell us whether either explanation is right.

Histograms tend to be more useful with larger data sets rather than smaller ones. Since drawing histograms can be tedious, the next few problems have used a computer to draw the histograms for you. Each problem asks you to describe the shape of the distribution.

Here is one format for describing a histogram. (Other formats are possible, of course.)

1. *Orientation and Background.* Tell the cases, the variable, its units of measurement, and the range of values.
2. *Describe the Shape.* Discuss modes, symmetry and skewness, tails, gaps and outliers. Try to use the concrete language of the context (here, states and exports) in preference to the abstract statistical terms.
3. *Interpretation.* Suggest possible connections between the patterns and the context.

54. *Exports.* This histogram uses the 50 U.S. states as cases, and shows the distribution of exports per capita (in thousands). The states with the largest values are Alaska (5.5), Washington (5.4), and Louisiana (3.6). Describe the distribution and include possible explanations for the patterns.

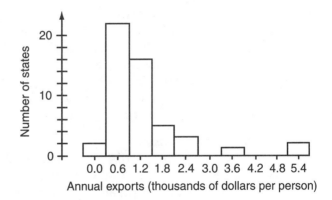

➤ *Solution*

For the 50 U.S. states , exports per person range from near $0 to more than $5000 (Alaska). Most of the states have per capita export values that form a single cluster between $0 and $2500, with values just under $1000 being most frequent. There are three outlying values, for states with unusually high exports in relation to population. Two are for Pacific coast states with low populations—Alaska and Washington—that export more than $5000 per person. It seems likely that for Alaska oil makes up the bulk of exports and for Washington lumber. The third state with atypically high exports is the Gulf state Louisiana, with $3600 per capita. The distribution for the remaining 47 states is roughly symmetrical. About 30 of the 50 states export between $500 and $1500 per capita each year.

55. You probably have some idea already about which states tend to have more farms. The histogram shows numbers of farms per million people, for the 50 U.S. states.

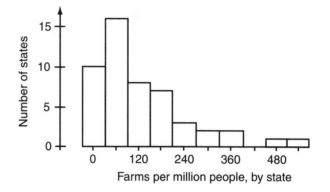

(a) List the aspects of shape you consider most noteworthy.

(b) Guess which 2 states have the highest numbers of farms per million people. (You'll need to think about population size as well as which states are farm states.)

(c) Write a short paragraph following the example as a model, describing the shape of the distribution and its possible meaning.

Answer

(a) The distribution has one mode, and is strongly skewed toward high values.

(b) North and South Dakota

(c) For the 50 U.S. states, the number of farms per million people ranges from a low of near zero to a high of more than 500. The distribution is strongly skewed toward high values, with more than half the states at 90 or below, and only six states having 270 or more farms per million people. Two states, North and South Dakota, are sparsely populated but depend heavily on agriculture. These two have the largest numbers of farms per million people, at roughly 500.

56. *Temperature Extremes.* What's the hottest it's ever been where you live? The coldest? Measured in degrees Fahrenheit, the difference between the record high and record low temperature can be as small as 88° or as large as 187°, depending on the state. The histogram takes the 50 U.S. states as cases, and shows the distribution of the difference between the temperature extremes.

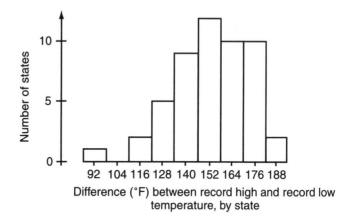

Difference (°F) between record high and record low temperature, by state

(a) List those aspects of the distribution you consider most noteworthy.

(b) Try to guess which 3 states correspond to the extremes, one at the low end, two at the high end.

(c) Write a paragraph describing the distribution.

Answer

(a) The distribution has only one mode, around 150° and is skewed toward low values. One state has a temperature difference about 20° below any of the others.

(b) Low end: Hawaii
High end: California, Nevada

(c) This histogram shows the difference between a state's record high and record low temperatures. For the states of the United States, these differences range from a low of 88°F to a high of 187°F, with 80% of the states (all but ten) almost uniformly distributed in the 50° range from 130°F to 180°F. The distribution is mildly skewed toward low values. The largest differences belong to California and Nevada, southwestern states whose deserts can be very hot and whose high Sierra mountains get very cold. The smallest difference belongs to the island state of Hawaii, whose temperatures are kept moderate by the Pacific Ocean.

57. *Hazardous Waste Sites.* Can you guess which states have the most, and the fewest hazardous waste sites? The histogram shows the number of sites per state for the 50 U.S. states.

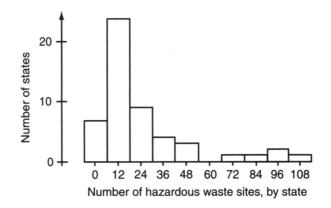

(a) List what you consider to be the main features of the shape of the distribution.

(b) Try to guess which states correspond to the extremes. (If you think you know which state has the most, you're probably right.)

(c) Write a paragraph based on your answers to parts (a) and (b).

Answer

(a) The distribution has one mode, a spike at 12, with almost half the states in the interval from 6 to 18. The distribution is strongly skewed toward high values, with a long right tail.

(b) New Jersey has the most sites.

(c) For the 50 U.S. states, the number of hazardous waste sites ranges from none to just over 100. Almost half the states have between 6 and 18 toxic sites; another 15% have fewer than 6. The distribution is strongly skewed toward high values, with 5 states in the range from 70 to a little over 100. Note that the values shown in the histogram give total numbers of sites, rather than sites per million square miles, or sites per million people. Adjusting for area or population would give a somewhat different picture.

58. *Drivers.* What percentage of the people where you live have a driver's license? The histogram takes the 50 U.S. states as cases and shows the distribution of the percentage of residents with a license.

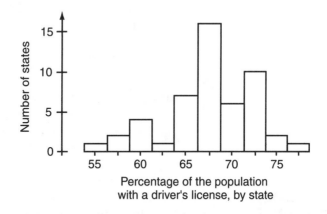

(a) List the main features of the shape of the distribution.
(b) List three factors that you think may influence whether a person has a license. Then use your list to make predictions: Which states do you expect with a high percentage, near 75%? A low percentage, near 55%?

Answer
(a) The distribution is lumpy, with almost half the states between 65% and 70%, and two smaller clusters of states, near 60% and in the low 70%s.
(b) Percentage of the population living in cities is one such factor. Urban states tend to have relatively fewer licensed drivers. Rural states have more.

59. *How Old is Wisconsin?* The University of Wisconsin has twelve campuses. Here are the years in which they were founded: 1848, 1866, 1868, 1871, 1874, 1894, 1909, 1916, 1956, 1968, and 1968.
(a) A stem plot shows a clear and striking pattern. Construct the plot, using paired stems (18•, t, f, s, 18∗, 19•, t, f, s, 19∗) and two-digit leaves.
(b) Briefly describe the pattern. (Write in terms of years and campuses, rather than just gaps and clusters.)

Answer

(a)
```
18•
  t
  f  |  48
  s  |  66 68 71 74
18*  |  94
19•  |  09 16
  t
  f  |  56
  s  |  68 68
19*  |
```

(b) One campus was founded before the Civil War (1848). Most were founded in the decades between the Civil War and World War I. After WWI, no new campuses were founded until the 1950s and 1960s.

2.4 FIVE-NUMBER SUMMARIES

One way to summarize a distribution is to tell the smallest and largest values (min and max) and also the three values that divide the data into quarters. The median divides the ordered data values in half, and the lower and upper quartiles divide these two halves in half again.

Taken together, the extremes (min, max), median (Q_2), and quartiles (Q_1, Q_3) give a handy summary called a five-number summary. You can show the summary in a graph called a box plot.

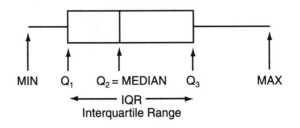

The Median

60. Find the medians for the two lists of numbers that follow.
    (a) 1, 2, 8
    (b) 1, 2, 8, 9

> **Solution**

(a) If you have an odd number of values, the median is the one in the middle.

1   2   8

median = 2

(b) If you have an even number of values, the median is the average of the two middle values, and falls exactly half way between them.

1   2   8   9

median = (2 + 8)/2 = 5

---

61. Find the median for the list of numbers 108, 115, 138, 139, 139.

Answer 138

62. Find the median for the list of number −3, 0, 2, 108.

Answer 1

The Quartiles

Different books use slightly different ways to define the quartiles. In practice, the difference doesn't matter, because all the definitions are based on the same idea—that the quartiles divide the data into quarters. Here are the two most common rules. Check with your textbook to find out which rule to use. (Don't waste time learning both!)

Median of the halves (Moore).

Q_1 = median of the lower half of the data
Q_2 = median of the upper half of the data

The next four examples show how to find the halves.

A.	1	2	3	|	4	5	6		n even, no ties
B.	1	3	3	|	3	7	8		n even, ties
C.	1	2	3	4	5	6	7		n odd, no ties, don't use the middle observation
				|					
				|					
				|					
D.	1	2	3	3	3	7	8		
				|					

Rounded Fractions (Johnson, Triola). Calculate the location (rank or depth) of the quartiles: Multiply the fraction ($\frac{1}{4}$ or $\frac{3}{4}$) times the number of values (n). If the result is a whole number, the quartile is the average of that value and the next one on the list. If the result is not a whole number, round *up* to the next whole number and take that value in the list.

(a) $n = 8$

Q_1: ($\frac{1}{4}$)(8) = 2. Q_1 = the average of the second and third values in the list.

Q_3: ($\frac{3}{4}$)(8) = 6. Q_3 = the average of the sixth and seventh values in the list.

(b) $n = 9$

Q_1: ($\frac{1}{4}$)(9) = $2\frac{1}{4}$. Round up to 3. Q_1 = the third value in the list.

Q_3: ($\frac{3}{4}$)(9) = $6\frac{1}{4}$. Round up to 7. Q_3 = the seventh value in the list.

63. Find the median and quartiles of the following two lists of numbers.

(a) 0, 1, 3, 5, 8, 10

(b) 0, 1, 3, 5, 8, 10, 20

➤ *Solution*

(a) The two middle values are 3 and 5, so the median is 4. We can find the quartiles using either method; both give the same values.

Median of the halves:

Q_1 = median of lower half

   = median of 0, 1, 3

   = 1

Q_3 = median of upper half

   = median of 5, 8, 10

   = 8

Rounded fraction:

$n = 6$

$\frac{1}{4}n = 1\frac{1}{2}$

Round up to 2

Q_1 = the second value in the list, or 1

$n = 6$

$\frac{3}{4}n = 4\frac{1}{2}$

Round up to 5

Q_3 = the fifth value in the list, or 8

(b) The median of seven values is the middle, or fourth, value. In this ordered list, it is 5.

Median of the halves:

Q_1 = median of lower half
    = median of 0, 1, 3 (don't include the median itself)
    = 1

Q_3 = median of upper half
    = median of 8, 10, 20
    = 10

Rounded fraction:

$n = 7$

$\frac{1}{4}n = 1\frac{3}{4}$

Round up to 2

Q_1 = the second value in the list, or 1

$n = 7$

$\frac{3}{4}n = 5\frac{1}{4}$

Round up to 6

Q_3 = the sixth value in the list, or 10

64–67. Find quartiles for each of the following data sets.

(a) **

3	123
4	458
5	
6	

($n = 6$)

(b) **

3	112
4	348
5	1
6	57

($n = 9$)

(c) **

3	112
4	348
5	111
6	579

($n = 12$)

(d) **

3	1127
4	348
5	11189
6	579

($n = 15$)

Answer

	MEDIAN OF THE HALVES		ROUNDED FRACTION	
	Q_1	Q_3	Q_1	Q_3
(a)	32	45	32	45
(b)	31.5	58	32	51
(c)	37.5	58	37.5	58
(d)	37	59	37	59

Box Plots

68. *Osmoregulation.* Find the five-number summaries and draw box plots for the two species of worms in Question 16.

➤ *Solution*

With six observations in a group, the median will be the average of the third and fourth in the ordered list; the quartiles are the second and fifth.

	N	G
Min	129.0	132.0
Q_1	129.0	158.0
Q_2 = median	145.0	177.5
Q_3	155.0	188.0
Max	156.0	191.0

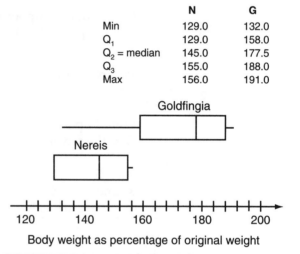

Body weight as percentage of original weight

It is useful to be able to locate the median and quartiles, at least approximately, from a plot.

69. *Stock Markets in Europe.* Question 18 asked you to draw two dot plots for percentage change for European stock markets.

(a) Use your dot plots to find five-number summaries for the two groups of markets.

(b) Use your summaries to draw parallel box plots for the two groups.

Answer
(a)

	Northern Europe	Southern Europe
min	−0.1	−25.6
Q_1	0.3	−7.9
Q_2	1.2	−2.7
Q_3	2.45	0.95
max	4.7	3.3

(b)

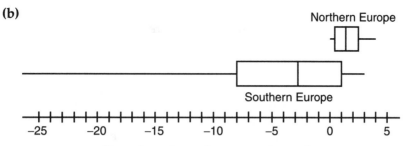

You can get an approximate five-number summary from a histogram by counting cases.

70. *Temperature Extremes.* Question 56 gives a histogram showing the distribution of the difference between record high and record low temperatures for the 50 U.S. states.
(a) Find an approximate five-number summary.
(b) Draw a box plot.
(c) Which features of the histogram show up in your box plot? Which are lost?

Answer
(a) The histogram is based on 51 cases (the 50 states plus the District of Columbia), so the median corresponds to the 26th largest value, and the quartiles correspond to the 13th largest and the 37th largest. The minimum (88) and maximum (187) were given in Question 56. If you use the scale on the y-axis to help you count cases, you find that Q_1 is the fifth largest of the values for the states in the bar at 140, and Q_3 is the largest of the values for the states in the bar at 164. Here are the values computed

from the raw data: min = 88, Q_1 = 142, Q_2 = 152, Q_3 = 168, max = 187.

(b)

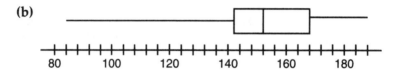

(c) The box plot's long left whisker tells us that the minimum value of 88 is much farther from the median than the maximum value is, but the box plot doesn't tell us much about the shape. For example, we can't tell how many modes there are; nor can we tell whether there is a long left tail, or just a single outlier.

Finding a five-number summary is easier if you already have a stem plot to work from, as in the next three problems. These problems, taken together, illustrate some of the uses and shortcomings of box plots: They are good for showing symmetry/skewness, center, and spread, but can't show clumping or modes.

71. *Mantle's Averages: Center and Spread.*
   (a) Use your stem plots from Question 38 to find five number summaries for Mickey Mantle's regular season and World Series batting averages.
   (b) Draw parallel box plots.
   (c) Use your two plots to compare the centers of the two distributions and the spreads of the two distributions.

Answer
(a) Mantle's batting averages

	Regular Season	World Series
min	.237	.120
Q_1	.275	.184
Q_2	.302	.229
Q_3	.314	.298
max	.365	.400

(b)

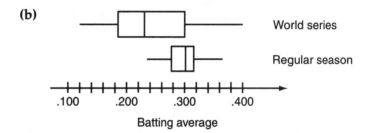

(c) Centers: Regular season averages tend to be higher.
Spreads: World Series averages are more variable.

72. *Salty Chips: Skewness and Symmetry.*
    (a) Use your stem plots from Question 39 to find five-number summaries for the tortilla chip data.
    (b) Draw parallel box plots.
    (c) Which distribution is the more skewed of the two? How can you tell?

Answer
(a) Salt content of chips (mg/oz)

	Rated Good	Rated Excellent
min	0	38
Q_1	40.5	67.5
Q_2	92	80
Q_3	145	128.5
max	198	170

(b)

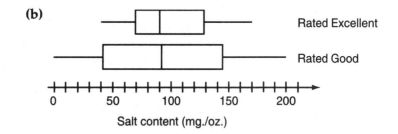

(c) This more skewed distribution is the one for the chips rated Excellent. The distance from Q_2 to Q_3 is greater than that from Q_1 to Q_2, and the distance from Q_3 to the maximum is greater than the distance from the minimum to Q_1.

73. *On Wisconsin: Bye-Bye Bimodality.* The twelve campuses of the University of Wisconsin were founded in years that form two clusters. A stem plot shows a clearly bimodal distribution.

(a) Compute a five-number summary for the Wisconsin data (Question 59).

(b) Use the five-number summary to draw a box plot.

(c) Why don't the two modes show up in your box plot?

Answer

(a) Year founded

min	1848
Q_1	1868
Q_2	1894
Q_3	1956
max	1968

(b)

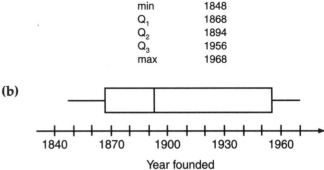

Year founded

(c) Bimodality is a feature of shape that cannot be read from any of the usual numerical summaries for a distribution. Because a box plot shows only the five-number summary, it can't show bimodality.

Modified Box Plots

Interquartile range. The distance between the upper and lower quartiles $(Q_3 - Q_1)$ is called the interquartile range, or IQR, and measures the spread of the middle 50% of the data.

Test for outliers. There's an informal test for outliers based on the IQR: Any value more than 1.5 IQRs from the nearest quartile is considered an outlier.

Modified box plots. Modified box plots are based on the outlier rule. They are like simple box plots but with two differences. First, instead of extending all the way to the smallest and largest values, the "whiskers" extend only as far as the smallest and largest values that are not outliers. Second, outliers are shown as isolated points. Here's an example to illustrate the outlier test and modified box plot.

74. *Stock Markets: Is Turkey an Outlier?*

(a) Refer to Question 18. Compute a five-number summary for the stock market data for the group that includes a Turkey.

(b) Use the IQR rule: Is Turkey an outlier?

(c) Draw a modified box plot.

➤ **Solution**

(a) There are eight observations in the group. The median is the average of the fourth and fifth: $[(-3.1) + (-2.3)]/2 = -2.7$. The two quartile rules give the same answers here. The first quartile is the average of the second and third smallest: $[(-10.4) + 9-5.4)]/2 = -7.9$. The third quartile is the average of the sixth and seventh: $[(-0.7) + 2.6]/2 = 0.95$, which we can round to 1.0.

$\min = -25.6$, $Q_1 = -7.9$, $Q_2 = -2.7$, $Q_3 = 1.0$, $\max = 3.3$

(b) $IQR = Q_3 - Q_1 = 1.0 - (-7.9) = 8.9$

$1.5\ IQR = (1.5)(8.9) = 13.35$, which we round to 13.4

Outlier check: Is Turkey (-25.6) more than 1.5 IQRs (13.4) from the nearest quartile (-7.9)?

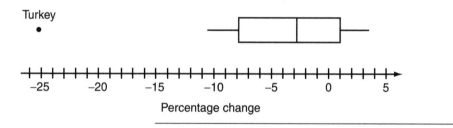

Turkey

−25   −20   −15   −10   −5   0   5

Percentage change

75. *Hotel Costs: Is New York an Outlier?*

(a) Compute a five-number summary for the hotel cost data of Question 36.

(b) Use the 1.5 IQR rule: Is New York an outlier?

Answer

(a) Hotel costs ($)

min	65
Q_1	69
Q_2	83.5
Q_3	97
max	129

(b) $IQR = Q_3 - Q_1 = 97 - 69 = 28$
$1 - 5\ IQR = 42$
$Q_3 + 1.5\ IQR = 97 + 28 = 125$
$125 < 129$

New York is an outlier

76. *Complaints: Is TWA an Outlier?* TWA scored worst out of ten airlines, both in terms of on-time percentage and in terms of complaints per 100,000 customers. For this problem, focus on the complaints.
(a) Compute a five-number summary.
(b) Use the 1.5 IQR rule: Is TWA an outlier?
(c) Southwest Airlines had an unusually low rate of complaints, only 0.19 per 100,000 customers. The next best, for Alaska Airlines, was 0.51, a rate more than two and a half times higher. Use the 1.5 IQR rule: Is Southwest an outlier?

Answer
(a) Complaints per 100,000

min	0.19
Q_1	0.61
Q_2	0.795
Q_3	1.16
max	1.31

(b) $IQR = 1.16 - 0.61 = 0.55$
$Q_3 + 1.5\ IQR = 1.16 + (1.5)(0.55) = 1.985$
$1.31 < 1.985$

TWA is not an outlier.

(c) $Q_1 - 1.5\ IQR = 0.61 - (1.5)(0.55) < 0$

Southwest is not an outlier.

77. *On-Time Percentage. Is TWA an outlier?* Now work with on-time percentages for the ten airlines.
(a) Compute a five-number summary.
(b) Use the 1.5 IQR rule: Is TWA an outlier? Are any other airlines outliers?
(c) Draw a modified box plot.

Answer

(a) On-time percentage

min	68%
Q_1	78%
Q_2	81.5%
Q_3	84%
max	89%

(b) IQR = 84% − 78% = 6%

1.5 IQR = 9%

Q_3 + 1.5 IQR = 93%

89% < 93%

Alaska Airlines is not an outlier.

Q_1 − 1.5 IQR = 78% − 9% = 69%

68% < 69%

TWA is an outlier.

(c)

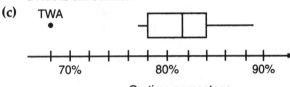

On-time percentage

2.5 THE MEAN AND STANDARD DEVIATION

The Mean

In a very general sense, the mean and standard deviation (SD) serve the same purposes as the median and interquartile range: they locate the center and measure the spread of a distribution. There are important differences, however. The mean and SD have special properties that make them more suitable than the median and IQR in some situations, but less suitable in others.

The next few pages of examples and problems first show how to compute the mean and SD, and then illustrate their properties and how they differ from the median and IQR. Because computing means and (especially) SDs for real data tends to be both time-consuming and boring, the problems that follow rely on made-up data. Later problems, in the next chapter, for example, will be based on real data.

The mean equals the sum of the observations, divided by how many there are.

78. Compute the mean of 0, 1, 2.

Answer 1

79. Compute the mean of 0, 2, 4, 6, 8.

> ***Answer*** 4

Because the mean = total/n for the data, the mean is often a good summary if your goal is to estimate an unknown total. For example, the mean crop yield for a sample of farms could be used to estimate total crop yield for a country. The mean selling price for a well-chosen sample of houses could be used to estimate the total market value of real estate in an area. Many sports averages are means. A basketball player's free throw percentage is the total number of free throws made divided by the number of attempts—a mean of observations that are all 1s (scores) and 0s (misses).

The deviations from the mean always add to zero.

80. For the data 1, 3, 5, 7, compute the mean, the deviations, and their sum.

> ***Solution***
>
> $$\text{Mean:} = (1 + 3 + 5 + 7)/4 = 4$$
> $$\text{Deviations:} = 1 - 4 = -3$$
> $$3 - 4 = -1$$
> $$5 - 4 = 1$$
> $$7 - 4 = 3$$
> $$\text{Sum of deviations} = (-3) + (-1) + 1 + 3 = 0$$

81. For the data 0, 1, 8, compute the mean, the deviations and their sum.

> ***Answer*** Mean = 3, Deviations are –3, –2, and 5. Their sum is 0.

If you think of the observations as dots on a number line, the deviations tell the distances from the dots to the mean. The mathematical property that the distances add to zero has a physical/visual meaning: the mean is the point where the dot graph will balance.

82. For the data 1, 3, 5, 7, draw a dot graph, and show the mean by putting a caret (^) under its location on the number line.

> ***Answer***

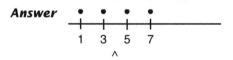

For each data set below, draw a dot graph and show the mean by putting a caret (^) under its location on the number line.

83. 0, 1, 2

Answer

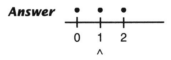

84. 0, 2, 4, 6, 8

Answer

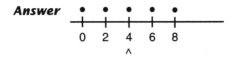

85. 0, 1, 8

Answer

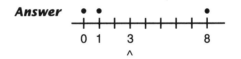

Because the mean equals the balance point, you can find the mean of any symmetric data set or distribution without arithmetic just by finding the point of symmetry.

Find the mean for each data set, using symmetry instead of computing.

86. 0, 100, 200

Answer 100

87. 1, 11, 21

Answer 11

88. 95, 100, 200, 205

Answer 150

89. 17, 20, 23, 26, 29

Answer 23

The Mean and Median Compared

A very important property of the mean, closely related to its being the balance point, is that the mean is influenced by extreme values (outliers and values far in the tail of a distribution). The median, by contrast, is robust; that is, it is not influenced by extreme values.

Compute the mean and median for the following sequence of data sets. (Note that there is a pattern to the data sets. Look for a pattern in your answers.)

90. 0, 1, 2

Answer Mean: 1; Median: 1

91. 0, 1, 5

Answer Mean: 2; Median: 1

92. 0, 1, 8

Answer Mean: 3; Median: 1

93. 0, 1, 98

Answer Mean: 33; Median: 1

94. 0, 1, 998

Answer Mean: 333; Median: 1

The mean's sensitivity to outliers has an important practical consequence. If your goal is to locate the center of a distribution, the mean tends to be a good choice if the distribution has one mode, is roughly symmetric, and has no outliers or very long tails. For skewed distributions or distributions with outliers or long tails, the median tends to be a better choice. (However, if your goal is to estimate an unknown total, the mean may be better than the median.)

Distributions that are roughly bell-shaped (one mode, symmetric, light tails, no outliers) are the sort that the mean was invented to summarize. Such distributions are so important that they have a chapter of their own, where there are many examples and problems with real data.

The Standard Deviation

If you use the median to locate the center of a distribution, then the interquartile range is usually a good choice for measuring the spread of the distribution. On the other hand, if your distribution is one that makes the mean a good choice for locating the center, you can measure spread using the standard deviation (SD).

The name gives a hint about the purpose of the SD: to measure the typical size ("standard") of a deviation from the mean.

The SD equals the root mean square of the deviation. To use this definition, read it backwards.

Deviations: Compute the deviations from the mean.
    Square: Square each one.
        Mean: Add them up and divide by $n - 1$.
          Root: Take the square root.

95. Compute the SD of 0, 1, 2.

➤ *Solution*
    Deviations: −1, 0, 1
      Square: 1, 0, 1
        Mean: $(1 + 0 + 1)/(3 - 1) = 2/2 = 1$
        Root: $\sqrt{1} = 1$

Compute the SD of each of the following data sets.

96. 10, 12, 14

    Answer 2

97. 0, 0, 3, 6, 6

    Answer 3

98. 1, 3, 5, 7, 9

Answer $2\sqrt{5/2} \approx 3.16$

99. 1, 3, 4, 5, 7

Answer $\sqrt{5} \approx 2.24$

The SD is the first of four or five difficult but crucial concepts in beginning statistics. It takes much more time and effort to understand than the IQR. One path to understanding is through the normal (bell-shaped) distribution, which is the subject of the next chapter. A second path is through a set of properties that tell how the SD behaves. That's what the next several problems are about.

Properties of the SD

A. The SD = 0 if there is no spread, no variation.

100.   Find the SD for the data set 10, 10, 10, 10, 10.

Answer 0

B. The SD increases as the spread increases.

Find the SDs of the following data sets. (There is a pattern that keeps arithmetic to a minimum.)

101.   0, 1, 2

Answer 1

102.   0, 2, 4

Answer 2

103.   0, 3, 6

Answer 3

104.   0, 5, 10

   Answer 5

For each pair of data sets below, decide which one has the larger SD. (Don't compute SDs.)

105.   0, 1, 2 and 0, 1, 3

   Answer Second SD is larger.

106.   1, 2, 2, 3 and 1, 2, 3, 4

   Answer Second SD is larger.

107.   1, 2, 3, 3, 4, 5 and 1, 3, 5

   Answer Second SD is larger.

108.   1, 3, 3, 3 and 1, 1, 3, 3

   Answer Second SD is larger.

   C. If each value in a data set is multiplied by the same number c, the SD gets multiplied by the absolute value of c. (The absolute value of 3 is 3, of –3 is also 3. Just drop a minus sign if it's there. Notation: $|-3| = 3$.)

Each data set below equals 0, 1, 2 multiplied by some constant. First find the constant. Then use property C to predict the SD. Then compute the SD, partly to verify your prediction, but also to help build your intuition about SDs.

109.   0, 5, 10

   Answer 5

110.   0, 2, 4

   Answer 2

111. 0, 50, 100

Answer 50

112. 0, −10, −20
Answer 10

The data set 1, 1, 3, 5, 5 has SD = 2. Each data set below comes from 1, 1, 3, 5, 5 through multiplying by a constant. Tell the SD without actually computing it.

113. 10, 10, 30, 50, 50

Answer 20

114. 2, 2, 6, 10, 10

Answer 4

115. .1, .1, .3, .5, .5

Answer 0.2

116. −25, −25, −15, −5, −5

Answer 10

D. If you add the same number d to each value in a data set, you don't change the spread, and the SD is unchanged.

Use dot plots to verify that the two data sets in each pair below have the same spreads. Then verify that the deviations from the mean are the same for both data sets in the pair. (If the deviations are the same, their root mean squares must be equal.)

117. 0, 1, 2 and 100, 101, 102

Answer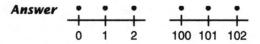

118. 1, 4, 6 and –2, 1, 3

Answer

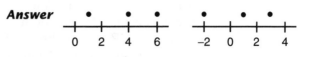

119. 21, 22, 30 and 0, 1, 9.

Answer

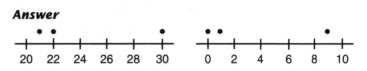

Each of the data sets in the problems below comes from 0, 1, 2 in two steps: multiplying by a number c, then adding some other number d to each value. Find the SD by finding c.

120. 100, 110, 120

Answer 10

121. –18, –17, –16

Answer 1

122. 19, 19.05, 19.1

Answer 0.05

For each pair of data sets below, guess the SD of the second from the SD of the first.

123. 0, 1, 2, 3, 4 (SD = $\sqrt{5/2}$) and 56, 58, 60, 62, 64

Answer $2\sqrt{5/2}$

124. 7, 8, 8, 9 (SD = $\sqrt{2/3}$) and –12.3, –12.6, –12.6, –12.9

Answer $.03\sqrt{2/3}$

125. 0, 1, 8 (SD = $\sqrt{19}$) and 21, 11, –59

Answer $10\sqrt{19}$

The SD and IQR Compared

The SD is very much influenced by extreme values in a way that the IQR is not. (When you have outliers or long-tailed distributions, the IQR tends to be a better measure of spread *if* your goal is to summarize the distribution.)

First, check that all distributions below have the same median and IQR. Compute the mean and SD for each, and notice the effect of the outliers.

126.   (a) −2, −1, −1, 0, 0, 0, 0, 1, 1, 2
(b) −3, −1, −1, 0, 0, 0, 0, 1, 1, 3
(c) −5, −1, −1, 0, 0, 0, 0, 1, 1, 5
(d) −20, −1, −1, 0, 0, 0, 0, 1, 1, 20
(e) −100, −1, −1, 0, 0, 0, 0, 1, 1, 100

Answer

All five distributions are symmetric, with mean = median = 0, $Q_1 = -1$, $Q_3 = 1$, IQR =2.

The SDs are
(a) $\sqrt{12/9} \approx 1.155$
(b) $\sqrt{22/9} \approx 1.563$
(c) $\sqrt{54/9} \approx 2.449$
(d) $\sqrt{804/9} \approx 9.452$
(e) $\sqrt{20004/9} \approx 47.145$

127.   If you know calculus, find the limit of the SD as c approaches infinity for the data set $-c$, −1, −1, 0, 0, 0, 0, 1, 1, c.

Answer

The SD is $\sqrt{(4 + 2c^2)/9}$. As $c \longrightarrow \infty$, this quantity increases without bound.

If the IQR is easier to understand than the SD and it is not influenced by outliers the way the SD is, why not just use the IQR all the time? Why bother with the SD at all? There are several possible answers. Here are just two. First, the IQR does not have the property that increasing the spread of a distribution always increases the IQR. (You just saw that in the last problem.) Second, and in my opinion, most important of all the reasons, the SD has a special property that makes it useful for measuring variability for averages: If you know the SD for a distribution and you take a random sample from the distribution, there's a simple formula for the SD of the average. (This comes in a later chapter.) There's no simple formula for the IQR of the average.

ADDITIONAL PROBLEMS

128. (Based on a 1991 project by Jeff Petry and Todd Houston.) Two students popped some Orville Redenbacher popcorn and recorded the percentage of kernels that popped within 2 minutes, 45 seconds. Here are their data:

Percentage

87.3%
82.3%
89.3%
82.4%
81.9%
88.8%
82.1%
81.0%
83.0%
85.6%

(a) Construct a stem-leaf diagram of these data.
(b) Find the median, the first quartile, and the third quartile.
(c) Using the $1.5 \times$ IQR criterion to determine outliers, how high a percentage of kernels must pop for the observation to be an outlier?
(d) Construct a box plot for these data.
(e) The sample mean is 84.4% and the sample standard deviation is 3.1%. Suppose we replaced each value by the percentage of kernels that did not pop, rather than the percentage that popped. What would be the new values of the sample mean and the sample standard deviation?

Answer
(a) Here is a stem-leaf diagram of the data.

81	0 9
82	1 3 4
83	0
84	
85	6
86	
87	3
88	8
89	3

Key: 87 | 3 = 87.3%

(b) There are 10 observations, so the median is the average of the 5th and 6th largest values, or (82.4% + 83.0%)/2 = 82.7%. The first quartile is the 3rd value in the ordered list, 82.1%, and the third quartile is the 8th value in the ordered list, 87.3%.

(c) The IQR is 87.3% − 82.7% = 4.6%. Thus, to be an outlier an observation would have to exceed 87.3% + 1.5 × 4.6% = 87.3% + 6.9% = 94.2%.

(d) Here is a boxplot of the data:

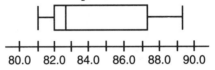

(e) The new variable is 100% −X. Thus, the new mean is 100% −84.4% = 15.6%. The new standard deviation is 3.1%, since the additive constant of 100 does not affect the SD and the SD of −X is the same as the SD of X.

129. The Registrar's Office collected data for the 628 new students who entered Oberlin College in the fall of 1994. One variable, measured Y, was the number of natural science credits earned by each student by the end of their first year at Oberlin. The average student had 7.5 credits; the SD was 6 credits.

(a) Make a rough sketch of the histogram for this variable, paying particular attention to the shape (which you should be able to infer from the average and SD given). Be sure to indicate the units on the X-axis (i.e., provide a scale for the histogram).

(b) The mean is
    (i) less than;
    (ii) equal to;
    (iii) greater than
the median. Choose one of these and explain your choice.

Answer

(a) Here is a histogram of the data. By knowing that the mean is 7.5 and the SD is 6, and knowing that it is not possible for a value to be less than zero, we can tell that the distribution is skewed to the right.

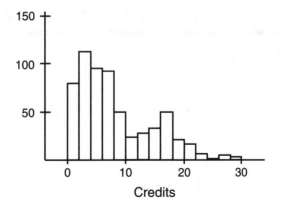

(b) The mean is greater than the median. When a distribution is skewed to the right, as this one is, the mean is greater than the median because the long right-hand tail pulls the mean up.

Data Appendix 1: AAUP Salary Data

Average 1984 Academic Salary, Percent Female, and Labor-Market
Conditions for Academic Disciplines

Discipline	Average Academic Salary	% Female	% Unemployed	% Nonacademic Jobs	Median Nonacademic Salary
Dentistry	$44, 214	15.7%	.1%	99.4%	$40,005
Medicine	43,160	25.5	.2	96.0	50,005
Law	40,670	34.0	.5	99.3	30,518
Agriculture	36,879	12.9	.8	43.4	31,063
Engineering	35,694	4.6	.5	65.5	35,133
Geology	33,206	13.5	.3	58.1	33,602
Chemistry	33,069	16.2	1.1	61.9	32,489
Physics	32,925	7.2	1.2	40.7	33,434
Life sciences	32,605	29.8	1.4	27.4	30,500
Economics	32,179	14.8	.3	34.2	37,052
Philosophy	31,430	23.1	1.8	17.1	18,500
History	31,276	30.5	1.5	20.5	21,113
Business	30,753	27.1	1.9	98.9	20,244
Architecture	30,377	31.6	1.1	98.3	21,758
Psychology	29,894	45.5	1.1	51.0	30,807
Educational psyc.	29,675	49.5	.9	82.5	20,195
Social work	29,208	80.0	1.9	98.6	16,965
Mathematics	29,128	15.4	.4	17.9	32,537
Education	28,952	48.1	.7	97.1	19,465
Sociology/Anthro.	27,633	40.9	2.1	12.8	21,600
Art	27,198	57.6	2.1	96.6	11,586
Music	26,548	45.5	4.3	98.5	16,193
Journalism	25,950	52.3	3.1	93.4	20,135
English	25,892	53.6	1.9	12.1	18,000
Foreign languages	25,566	55.0	3.5	12.1	20,352
Nursing	24,924	94.2	1.4	96.2	17,505
Drama	24,865	58.5	5.5	97.1	20,005
Library science	23,658	82.2	1.1	78.9	15,980

SOURCE: Bellas, Marcia and Barbara F. Reskin, "On Comparable Worth," *Academe*, September–October 1984, p. 84.

Data Appendix 2: Westvaco Data

Variables:

Exempt: 1 = salaried employee

0 = hourly employee

RIF (Reduction in force)

0 = retained

1 = laid off in Round 1 of reductions

2 = laid off in Round 2 of reductions

etc.

ID	Job Title	Exempt	BIRTH mo	yr	HIRE mo	yr	RIF	Age 1/1/91
2	Supv Engineer Serv	1	4	54	6	72	0	37
4	Design Engineer	1	12	43	9	67	0	48
6	Proj Eng-Elec	1	9	43	4	71	0	48
7	Engineering Manager	1	2	32	11	63	0	59
8	Packaging Engineer	1	3	38	11	83	0	53
11	Machine Designer	1	9	59	3	90	0	32
12	Supv Machine Shop	1	11	37	3	64	0	54
14	Design Engineer	1	3	37	6	74	0	54
15	Project Engineer	1	6	34	8	81	0	57
16	Prod Sped-Printing	1	12	44	11	74	0	47
17	Design Engineer	1	1	31	3	67	0	60
26	Engineering Associate	1	2	57	4	85	0	34
28	Customer Serv Engr Assoc	1	2	62	5	88	0	29
29	Project Engineer	1	7	49	9	73	0	42
31	Engineering Assistant	1	6	60	7	86	0	31
32	Project Engineer	1	8	43	4	64	0	48
35	Customer Serv Engineer	1	4	30	9	66	0	61
36	Design Engineer	1	3	36	2	78	0	55
38	Engineering Clerk	0	9	66	7	89	0	25
41	Secretary to Engin Manag	0	2	43	9	66	0	48
43	Engineering Tech II	0	4	53	8	78	0	38
47	Engineering Tech II	0	10	35	7	65	0	56
1	Prod Specialist	1	9	27	10	43	1	64
5	Machine Designer	1	2	39	4	85	1	52
20	Design Engineer	1	9	38	12	87	1	53
21	Project Engineer	1	7	25	9	59	1	66
25	Chemist	1	8	22	4	54	1	69
27	Engineering Associate	1	2	61	9	85	1	30
30	Machine Parts Cont-Supv	1	10	28	8	53	1	63
39	Parts Crib Attendant	0	11	69	10	89	1	22
45	Engineering Tech II	0	8	36	3	60	1	55
46	Engineering Tech II	0	8	38	9	74	1	53
50	Engineering Tech II	0	1	32	2	63	1	59
9	VH Prod Specialist	1	5	35	9	55	2	56
10	Electrical Engineer	1	11	49	3	86	2	42
18	Design Engineer	1	4	60	5	89	2	31

(Data Appendix 2 continued)

	ID Job Title	Exempt	BIRTH		HIRE		Age	
			mo	yr	mo	yr	RIF	1/1/91
19	Chemist	1	12	30	10	52	2	61
23	Machine Parts Cont Coor	1	9	37	10	67	2	54
33	Machine Designer	1	3	35	12	68	2	56
44	Engineering Tech II	0	5	36	4	77	2	55
48	Engineering Tech II	0	8	27	12	51	2	64
49	Technical Secretary	0	5	36	11	73	2	55
22	Printing Coordinator	1	2	41	1	62	3	50
24	Prod Dev Engineer	1	6	59	11	85	3	32
40	Engineering Tech II	0	2	36	4	62	3	55
3	Prod Specialist	1	7	32	1	55	4	59
13	VH Prod Specialist	1	3	42	4	62	4	49
37	Engineering Tech II	0	9	58	11	76	4	33
42	Engineering Tech II	0	7	56	5	77	4	35
34	Engineering Associate	1	8	68	5	89	5	23

SOURCE: Robert A. Martin v. Envelope Division of Westvaco Corporation, 92-3121-F.

$\overline{x}$

3	276.7	WaPost h	4.80	17	92	335.50	−2.13
13	75	WaWater	1.24	14	326	18.50	−.13
12.13	7.50	WastMln	..	..	1417	7.63	−.13
33.63	18.38	Waters	..	51	501	30.38	+.50
43.75	17.00	WatkJn	.48	73	433	26.25	−.25
34.38	11.31	Watsco s	.14	33	116	29.50	−.63
24.63	15.50	Watts h	.31	..	363	24.38	+.25
6.63	1.00	Waxmn	..	3	786	5.88	−.13
38.13	23.13	WthfrdE	..	..	1012	35.63	−.13
20.00	15.25	WebbD	.20	..	274	16.50	..
36.50	23.50	Weeks n h	1.72	34	763	36.00	..
44.75	34.25	WeinRl	2.48	22	374	42.88	−.25
4.38	2.00	Weirt	..	..	901	3.00	..
34.88	28.75	WeisMk	.92	16	68	30.75	−.38
24.88	15.88	Wellmn	.32	20	666	17.63	−.25

The Normal Distribution

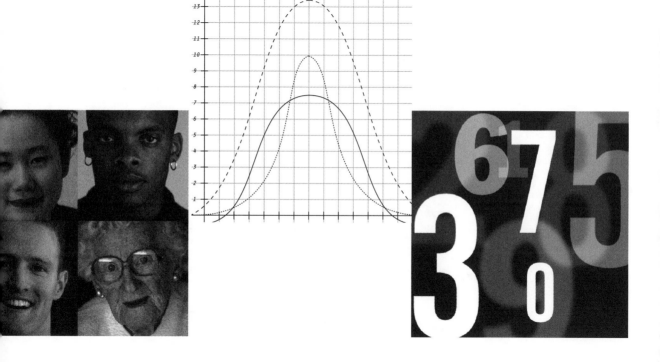

Summary

3.1 NORMAL DISTRIBUTIONS

We have seen how to describe data distributions in general in Topic 2, and now we turn to the important family of **normal distributions.** If we know the mean, μ, and the standard deviation, σ, for one of these distributions, we know everything about that distribution. Below you see several normal distributions. They

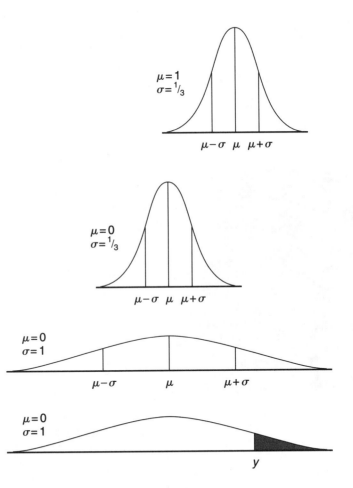

are all at the same scale, but they have different values for their means (their μ's) and standard deviations (their σ's). On each graph, the points $\mu - \sigma$, μ, and $\mu + \sigma$ are marked. Note that the value of μ and σ shown is that for that particular distribution and varies from graph to graph.

You will see that each of these distributions is symmetric about its mean, μ, and that each has a single mode, also at μ. This symmetry implies that the median of a normally distributed variable is equal to the mean, μ, as well. The points one standard deviation away from the mean, $\mu - \sigma$ and $\mu + \sigma$, mark the flex points for these curves. As you pass by these points, the graph above changes from curving down to curving up or vice versa.

How μ and σ Relate to the Graph

Looking at these examples you will see that changing μ while σ is unchanged shifts the curve to the right or the left, leaving its shape unchanged. Leaving μ unchanged while making σ smaller squeezes the curve inward horizontally toward μ and makes it bulge up vertically. When σ, the standard deviation, is smaller the values of the variable should be more concentrated about the mean, μ, and this is what you see. The area under these curves to the left of a given point y on the axis gives the probability of a variable with this distribution being less than y. So the area below the curve to the right of μ is always $1/2$ and the area under the whole curve is 1. We have used the same scales throughout, so if the curve is shrunk horizontally by a factor of $1/3$ toward μ, it must be stretched up by a factor of 3 to keep the total area equal to 1. This is what the graphs show.

Standard Units: z

Each normally distributed variable has its mean, μ, that gives its center and its standard variation, σ, that gives its spread. The graphs we have seen suggest that the values are concentrated near the center for a normal distribution and that σ is a natural unit for measuring the deviation of an observed value y of a normal variable from its center, μ. We use these ideas to define the distance, z, from y to μ in **standard units.** First, we note that y is distance $y - \mu$ to the right of μ. Next, we divide by σ to see how many standard deviations this is. The result gives us our z: $z = (y - \mu)/\sigma$, the value of y in **standard units.** You also find z referred to as the **standardized value** of y. Whatever the normal distribution of y, z is always normally distributed with $\mu = 0$ and $\sigma = 1$. The values found in tables of **standard normal probabilities,** which give the probability of a random standardized value being less than z for selected values of z. Remember that such probabilities are given by the area under the standard normal curve to the left of z. A short table of standard normal probabilities is shown below.

Brief Table of Standard Normal Probabilities

z	A(z) = area to the left of z
−3	0.0013
−2	0.0228
−1	0.1587
0	0.5000
1	0.8413
2	0.9772
3	0.9913

From this table we can calculate how the values of a normal variable cluster about its mean. Values within one standard deviation of the mean correspond to z-values between −1 and 1. The area under the curve and above this interval is the area to the left of 1 on the axis take away that to the left of −1. From our table this area is .841 − .159 = .682. By convention, we round this going to percent to get 68%. This is the first percentage in the **68/95/99.7 rule** for the percentage of values of a normally distributed variable that fall within 1, 2, or 3 standard deviations of the mean. The other values can also be worked out from our short table.

Similar calculations from tables of standard normal probabilities allow us to find the probability of a z-value falling within any interval bounded by the tabulated points. If the interval extended upward from without a limit, then we are looking for the area under the entire curve take away the area to the left of z. This is (1 − the tabulated value). For example, the probability of a z-score's being greater than 1 is 1 − .841 = .159, where .841 is the value found in our table for the z-value 1.

How do we find the probability that a normal variable y with $\mu = 20$ and $\sigma = 9$ will be observed to be between 15 and 23? We simply find the z-values corresponding to this range, using $z = (y − \mu)/\sigma$. The z-values are −.55 and .33, rounded down. Using a more complete table than ours, you would find the area to the left of .33 to be .6293 and that to the left of −.55 to be .2912. So the probability that y lies between 15 and 23 is .6293 − .2912 = .3381.

The examples that follow show how a table of normal probabilities can be used to solve problems of various sorts. These methods are not only of practical use in estimating the probability of events, but they are also fundamental in understanding methods of inference for normally distributed data.

With the standard normal curve we have a way of judging just how probable various observations are. This will later give a quantitative approach to choosing between hypotheses based on how likely they make our observations. Because many populations

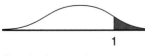

1

Graph of z: $\mu = 0, \sigma = 1$

and variables yield data that is roughly normal (or can be made so by a transformation), techniques based on the normal distribution are widely used.

Self-Testing Questions

1. Records of the "waiting times" between earthquakes (of magnitude 6 or more) in Parkfield, California, follow a normal curve with mean = 22 years, SD = 3 years. If the normal model is correct, then 10% of the waiting times will be more than 26 years. It follows that 90% of the waiting times will be more than how many years?

   Answer 18

2. The winning times for the Boston Marathon for the years 1957 through 1978 are approximately normal with mean =138 minutes and SD = 5 minutes.

   According to the approximation, what does the 25th percentile of the distribution equal, to the nearest whole number?

   ➤ ***Solution***

   $z = -.68$, so time = $138 - .68 \times 5 = 124.6$ or approximately 135. Out of these 21 years, 6 of the running times are less than or equal to 135.

3. Find the shaded areas in each graph of the standard normal curve.

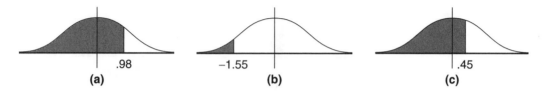

.98	−1.55	.45
(a)	**(b)**	**(c)**

Answer
(a) 83.65%
(b) 6.06%
(c) 67.36%

4. What are the cumulative probabilities associated with the following z-scores?
(a) 2.54
(b) 1.09
(c) −.51

Answer
(a) 99.45%
(b) 86.21%
(c) 30.50%

5. What are the z-scores associated with the following cumulative probabilities?
(a) 14.92%
(b) 99.38%

Answer
(a) −1.04
(b) 2.50

6. Find the probability associated with the shaded region in each graph of the standard normal curve.

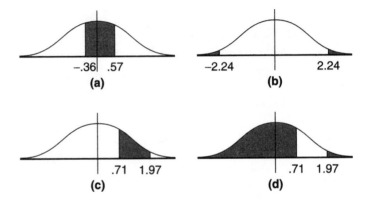

Answer
(a) .3562
(b) .0251
(c) .2145
(d) .7855

7. The heights, in inches, of a statistics class of 80 students is normally distributed with mean = 69 and SD = 3.5. How many students would we expect to be more than 6'2"?

➤ *Solution*

6'2" (74 inches) is 1.43 standard units away from the mean. The probability associated with this height is .0764. Thus, about 6 students would be over 6'2" in height, since 80 × .0764 = 6.1.

8. Given a normal distribution with mean = 16 and SD = 5, how much of the distribution will fall between 1 and 26?

➤ *Solution*

1 is 3 standard units away from the mean and 26 is two standard units away from the mean. Following the 68/95/99.7 rule, the probability within these two boundaries equals (.5)(95) + (.5)(99.7) = 97.35%.

9. The pulses of people in a certain population follow a normal distribution, with a mean of 70 and a standard deviation of 10. What percentage of this population has a pulse greater than 85?

➤ *Solution*

The z-score for 85 is $\frac{(85 - 70)}{10} = 1.5$. The area to the right of 1.5 under a z-curve is .0668, so about 6.7% of the people have pulsed greater than 85.

10. Consider the pulse distribution from Question 9. If a person is chosen at random from the population, what is the chance that the person will have a pulse between 60 and 85?

➤ *Solution*

The z-score for 60 is $(60-70)/10 = -1.0$. The z-score for 85 is $(85-70)/10 = 1.5$. Thus, we need to find the area between -1.0 and 1.5 under a z-curve. This area is $.9332 - .1587 = .7745$.

11. Consider the pulse distribution from Question 9. What is the 30th percentile of the distribution?

➤ *Solution*

The z-score (i.e., the standard units value) for the 30th percentile is $-.52$, so the equation we need to solve is $-.52 = (c-70)/10$. This means that c, the cutoff number we want, is $70 - .52 \times 10 = 70 - 5.2 = 64.8$.

12. The heights of American women follow a normal distribution, with a mean of 64 inches and a standard deviation of 2.5 inches. What percentage of American women have heights between 61.5 inches and 66.5 inches?

➤ *Solution*

The range 61.5 inches to 66.5 inches covers the mean plus or minus 1 standard deviation, so we can expect 68% of American women to have heights in this range.

13. The heights of American women follow a normal distribution, with a mean of 64 inches and a standard deviation of 2.5 inches. What is the 90th percentile of heights of American women?

➤ *Solution*

The z-score (i.e., the standard units value) for the 90th percentile is 1.28, so the equation we need to solve is $1.28 = (c-64)/2.5$. This means that c, the cut-off we want, is $64 + 1.28 \times 2.5 = 64 + 3.2 = 67.2$ inches.

14. The heights of American women follow a normal distribution, with a mean of 64 inches and a standard deviation of 2.5 inches. What is the probability that a randomly chosen American woman is more than 66 inches tall?

➤ *Solution*

The z-score for 66 inches is $^{(66-64)}/_{2.5} = 0.8$. The area to the left of 0.8 under a z-curve is .7881 and the area to the right is $1 - .7881 = .2119$. Thus, the probability is .2119, or about 21%.

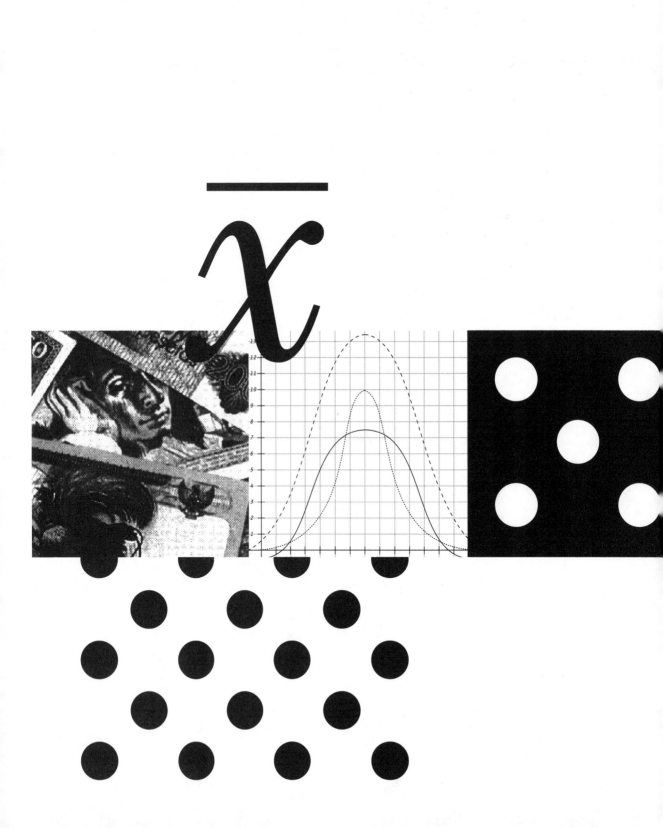

Sequence Data

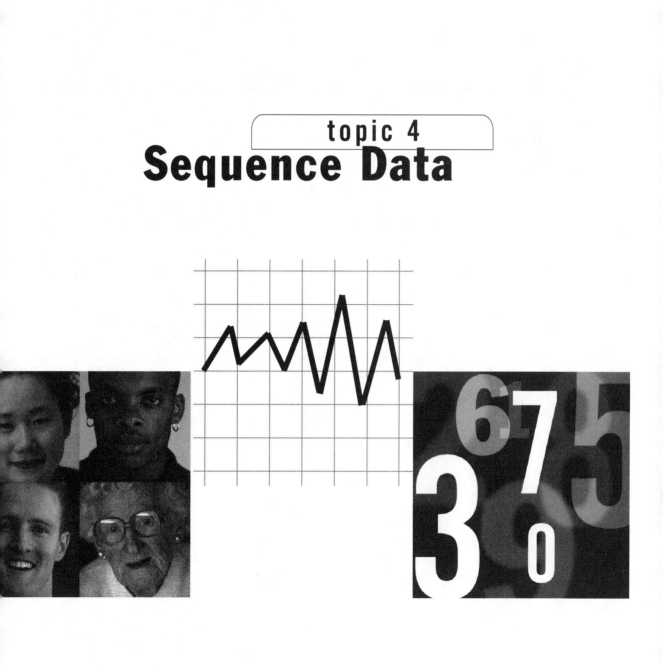

Summary _____

4.1 HOW ONE THING VARIES WITH ANOTHER

From ancient times the seasons were tracked by marking the rising or setting points for the Sun. Stonehenge is partly a means of recording the rising and setting of the Sun and thus recording **sequence data,** a series of values of a variable over time. Such data include the hourly temperatures on a patient's chart, the daily closings for the Dow Jones averages, the annual levels of sunspot activity. We plot such data with time shown on the horizontal axis and the other variable on the vertical axis. For example, if the Dow Jones closed at 5612 on August 30, we plot a data point at height 5612 above the position on the horizontal axis for August 30.

When sequence data is plotted, you can see how it varies over time. This is the simplest example of the variation of one variable with another. The statistical techniques used to study joint variation were developed in agriculture where researchers studied such things as how crop yield varied with irrigation. With many experimental plots and many levels of irrigation, the data produce a **scatterplot,** a cloud of points, rather than the simple sequence plots we consider here. We will look at these in Topic 5, but the ideas developed in looking at sequence data and their trends will help us understand scatter plots.

4.2 PATTERNS

We describe how a single variable changes over time by first identifying the **trend.** Second, we check the **deviations** from the trend. Common trend patterns are **linear growth, exponential growth,** or **logistic growth.** Two plots of data with these trends are shown on the top of page 95.

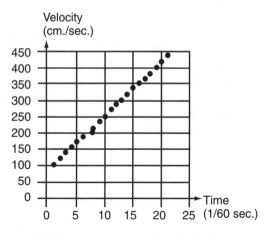

Freefall Showing Linear Growth: Velocity (cm./sec.) Versus Time (¹⁄₆₀th of a second)

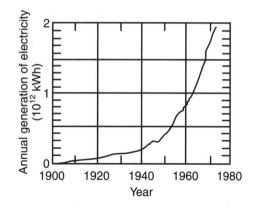

Exponential Growth: Annual Energy Used to Generate Electricity (U.S.)

To quantify treatment of sequence data, we need a formula for the trend line so that the deviations can be measured. The spirit can be understood by looking at the plot of fruit fly data. The ideal logistic curve has been drawn on the plot, and the data points show as circles. Looking at the plot, you see that the deviations are small, with no systematic pattern. These are minor chance variations from the logistic model. The model below fits the data well, and the deviations are random and without further pattern.

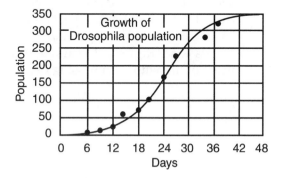

Logistic Growth: Fruit Fly Population Versus Time

Yearly records of Zurich sunspot number, **Z,** have been kept for almost three centuries. A graph of this data is shown below. There is no overall growth, so the trend line is horizontal and stands at the average for **Z,** somewhere around 40. The deviations from this trend show the cyclic pattern of the 11-year solar cycle. This pattern in the deviations shows there is additional structure in the data that should be accounted for. Our model needs to be expanded to take cyclic variations into account.

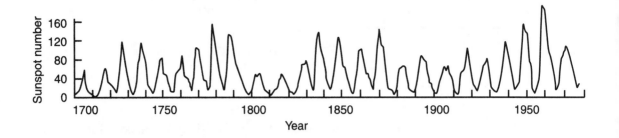

4.3 SLOPES AND GROWTH RATES

In **linear growth** the increase in the variable over a period depends only on the length of the period, independent of when the period begins. The amount added per unit time is a constant called the **growth rate.** You can visualize this rate as the **slope** of the trend line, with numerical value equal to Rise/Run =(change in y)/(change in x). The increase along the trend line over a time interval of length L is Slope $\times$ L. For example, the graph below shows how life expectancy has been increasing. Over the 40 years from 1950 to 1990, life expectancy increased from 68.2 years to 75.4 years, a Rise (change in y) of 7.2 years. The rate of increase equals Rise/Run = 7.2 years/40 years or about 1.8 additional years of life with each passing decade.

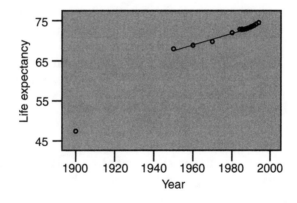

In **exponential growth** the increase in the variable over an interval of fixed length is a fixed multiple of the value at the start of the interval. For example, the energy used to generate electric power in the United States shows roughly exponential growth, increasing by 100% (doubling) about every decade. A **logarithmic transformation** of the variable helps us see how close this is to exponential growth. If growth were precisely exponential, the plot of the logarithms of the energy consumed would be straight. Instead, as you see below, the plot curves down, showing a decrease in growth rate with time. The wiggles show further fluctuations in the growth rate.

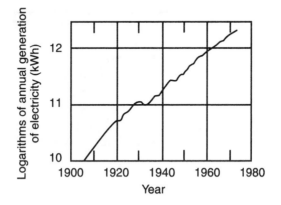

4.4 A NOTE OF CAUTION

Even the simplest and most direct relationships between variables can fail outside the tested ranges. Readings on a kitchen scale will be accurate for things weighing from a few ounces to a few pounds, but this linear relationship between weight and readings will fail for things that are too heavy or too light. Use caution when **extrapolating,** that is, applying a relationship outside its tested range. Also remember that relationships between variables may shift over time, either gradually or suddenly. This is like a scale going out of adjustment.

4.5 SEQUENCE DATA: WHAT TO LOOK FOR

Once the data is plotted, you should look for trends, especially for linear, exponential, or logistic growth. Next, look for the deviations from the trend. If these are relatively small chance variations, then your trend accounts for what is observed. Watch for cyclic or other patterns in the deviations. These show there are further aspects of the data that need to be considered.

Plotting the logarithm of the variable changes exponential growth to the easily recognized pattern of linear growth. This is one example of a transformation of the variable that helps in identifying the trend.

When a trend has been identified, quantitative methods can be used to fit the best trend line to the data. For example, the straight line that best matches the data in a scatterplot is the **regression line.** Its slope is one of several numerical summaries of the data that we will look at in Topic 5, "Exploring and Describing Relationships."

Self-Testing Questions

1. Make a sequence plot for the period 1850 to 1995 that shows the percentage of males in the United States. Use the following data. The percent of males stayed steady at 51% until 1910, when it went to 52%. The percentage dropped to 51% in 1920 and dropped again in 1940 to 50%. Since 1960, the percentage has stayed at 49%.

Answer

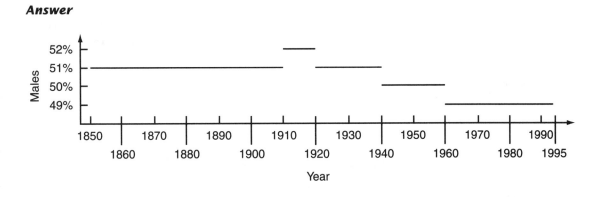

2. The population of Florida at different times is given in the following table. Make a sequence plot of these data.

Year	Population
1910	752,000
1920	968,000
1930	1,468,000
1940	1,897,000
1950	2,771,000
1960	4,952,000
1970	6,791,000
1980	9,747,000
1990	12,938,000

Answer

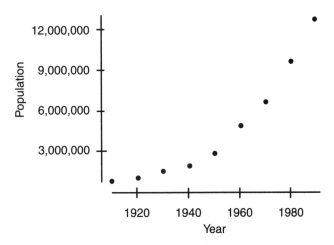

3. Describe the pattern in the plot from Question 2.

Answer The plot shows exponential growth.

4. The following plot shows the number of larcenies, in thousands, reported in the National Crime Survey for the years 1975–1977. (Month 1 is January 1975, month 13 is January 1976, etc.) What pattern best describes these data?

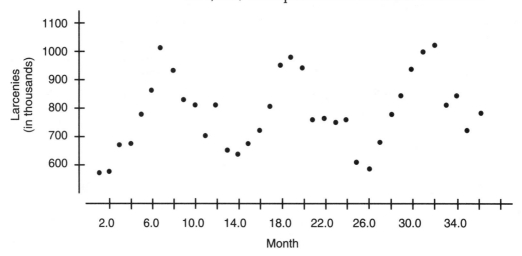

Answer The data shows a cyclical pattern.

5. The following graph shows how the death rate (deaths per 100,000 population) from cancer has changed over time. What pattern best describes these data?

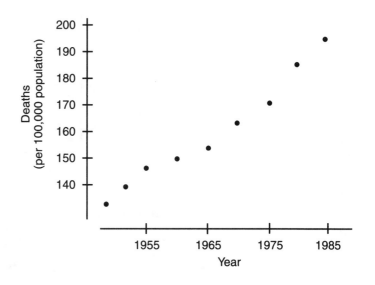

Answer

There appears to be a change point at 1965, since the slope is steeper after 1965 than it was before 1965.

6. Consider the plot from Question 5. Use the data for the years 1965, when the death rate was 153, and 1985, when the death rate was 193, to find the rate of increase in the death rate.

Answer

The slope is $^{193-153}/_{1985-1965} = ^{40}/_{20} = 2$. So the death rate went up by 2 per year during this time period.

7. The following graph shows how the number of Ph.D.s in mathematics awarded to U.S. citizens has changed over time. What pattern best describes these data?

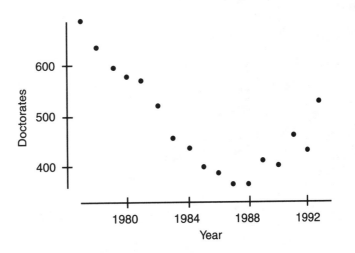

Answer

There appears to be a change point at 1987, since the slope is negative before 1987 but positive after 1987.

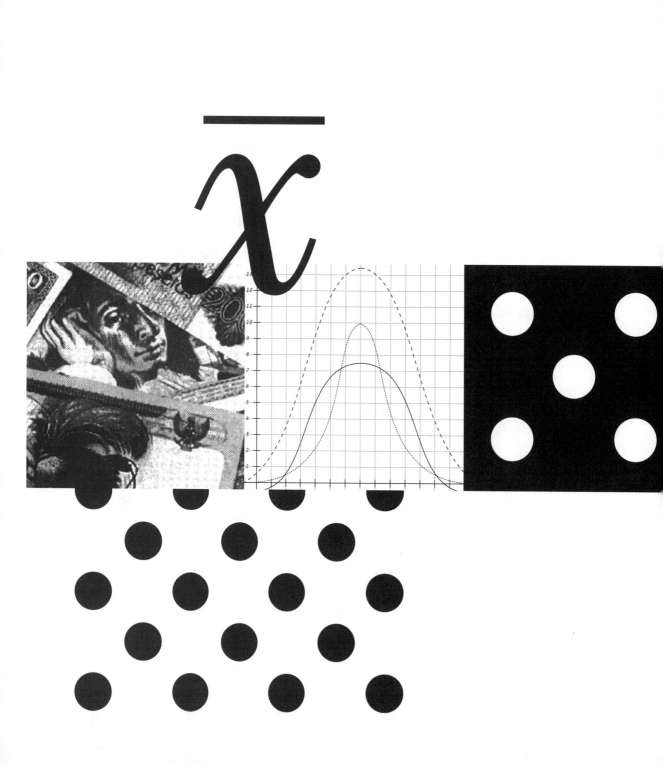

topic 5
Describing Relationships

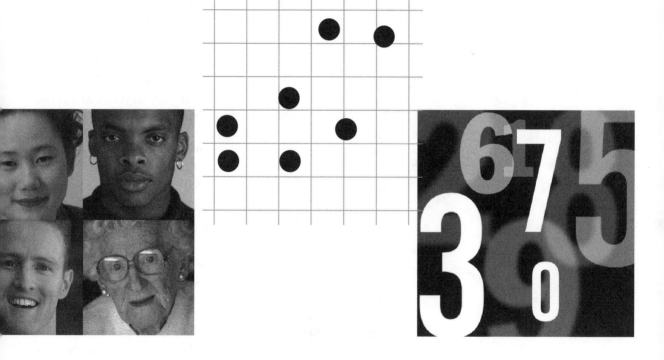

Summary _____

INTRODUCTION

When to review this material will depend on the book you are using. This workbook and the CD follow David Moore (*Introduction to the Practice of Statistics, Basic Practice of Statistics*) and others who organize the topics of a first course to fit the modern distinction between exploration-and-description and inference. This means that if your textbook follows a more traditional organization (e.g., Mario Triola's *Elementary Statistics* or Robert Johnson's *Elementary Statistics*), you'll want to skip over much of this chapter for now and return to it later in connection with inference for the situations covered here. Come back to Section 5.1 when you study regression and correlation; come back to Section 5.2 when you study analysis of variance; and come back to Section 5.3 when you study two-way tables and the chi-square test.

Where does this topic fit in? A good way to think about the ideas here is to relate them to a familiar pattern, the one you first met in Topic 2, Plot, Shape (Transform), Summaries. Here, as in all of statistics, numerical summaries only make sense for certain shapes. Whichever summaries you use, you can think of them as telling you about fit (= pattern) or about residuals (= deviation). Topic 2 told how to explore and describe patterns in the behavior of a single variable. Topic 4 introduced what for most people is the most familiar kind of two-variable relationship: changes over time in the values of a single variable. We now turn to exploring and describing patterns that relate any pair of variables. Often it makes sense to think of one of the two variables—the **predictor**—as coming before the other variable—the **response.** Sometimes the predictor is literally before in time, and sometimes it is before only in a logical sense. Since each of the predictors and responses can be either categorical or quantitative, there are four basic structures for two-variable relationships. The four numbered sections that follow correspond to these four structures.

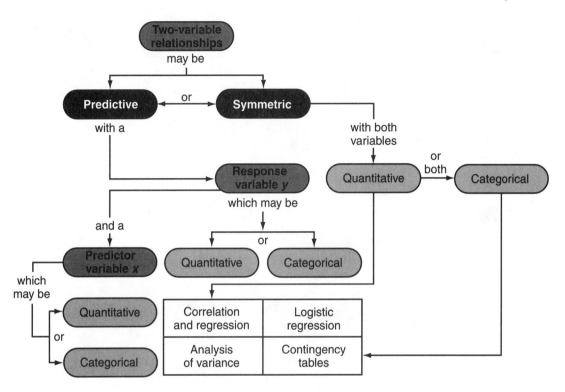

Describing Relationships

5.1 REGRESSION AND CORRELATION

Response and Predictor Are Both Quantitative

Where does this topic fit in? Regression and correlation deal with relationships between two quantitative variables. The main **plot** is a scatterplot; the ideal **shape** suggests an oval balloon. For such plots, you can give numerical **summaries** by fitting a line for predicting y-values from x-values (**fit**); you can also give a correlation coefficient that tells how fat or skinny the balloon is (**deviation** from fit).

Note: Your textbook probably doesn't cover balloon summaries, but they are quick to learn, and they give you a simple, visual way to review many of the important ideas of regression and correlation. Because books don't cover this useful and powerful way of thinking, a special section (Balloon Summaries: A Visual Approach to Regression and Correlation) is included at the end of the summaries for Topic 5.

Shape Remind yourself that a distribution's ideal shape is the normal one: symmetric, with one mode, light tails, and no outliers. For a scatterplot, the ideal shape is an oval balloon:

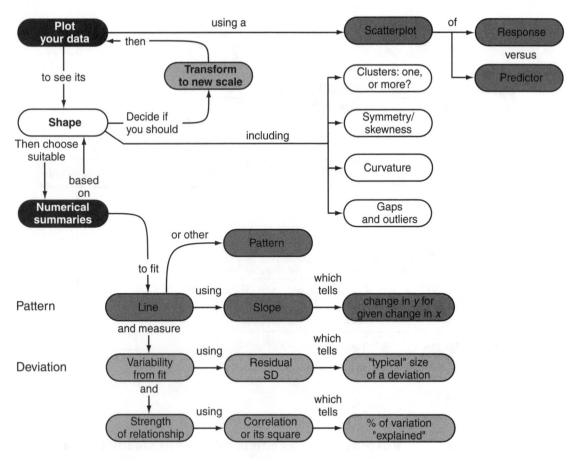

Describing Relationships: Both Variables Quantitative

symmetric, with one cluster of points, no curvature, and no out-liers. (If X and Y are both roughly normal, the scatterplot will tend to have a balloon shape.) Departures from the ideal shape often fall into one of four categories.

1. *Clusters.* Is there just one, or more than one? More than one suggests cases of two or more kinds.
2. *Curvature.* Is the relationship linear or curved? Often a curved relationship can be straightened by transforming either the response, the predictor, or both to new scales.
3. *Symmetry.* A balloon has two lines of symmetry, but some plots don't. For example, a wedge-shaped plot, narrow at the left and wide at the right (or vice-versa) indicates greater vari-ation (larger SD) when the predictor values are larger, and suggests transforming the response to a new scale.

4. *Isolated points*. Outliers are isolated points in the y (response) direction; influential points are isolated in the x direction. Such points can have a disproportionate influence on the values of fitted lines and summary numbers. You can gauge their effect by doing a second analysis without them and comparing it to your first analysis.

Fit The fitted line is determined by

1. The fitted slope $\hat{\beta}$ (or b), and
2. The "anchor point" $(\bar{x}, \bar{y})$.

The easiest way to remember the equation of the fitted line is to start with the definition of slope:

$$\text{Slope} = \frac{\text{Rise}}{\text{Run}} = \frac{\text{change in } y}{\text{change in } x}$$

The equation says that the fitted slope (which will be an actual number) equals "rise over run" for any point (x, y) and the anchor point $(\bar{x}, \bar{y})$.

The **principle of least squares** says to choose the fitted line that makes the sum of squared errors as small as possible. Which errors? Because we use x-values to find fitted y-values, the error we want to use is the residual = observed y - fitted y. These residuals correspond to the *vertical* deviations from the points to the fitted line.

Deviations Once we have a fitted line, we can use residuals to judge how well the line fits. Together, the sizes of the residuals tell us how much of the variation in y-values is left over after using x-values to find fitted y-values. You can also use a **residual plot** (of residuals versus fitted values) to check for patterns. (The CD includes four data sets created by Francis Anscombe of Yale University. All have exactly the same values of the usual summary statistics—the same intercept, slope, correlation coefficient, residual sum of squares, etc. However, each plot shows a very different pattern. The moral: *Always plot your data!*)

The **correlation coefficient** measures, on a scale from −1 to 1:

1. The direction (+ / −) and strength of any linear relationship between x and y,
2. The direction and shape (fat or skinny) of a balloon summary of the x, y scatterplot,

3. The slope of a fitted line when x and y are in standard units,
4. The percentage of variation in y that can be predicted from x using a fitted line, and
5. The size of the regression effect.

The correlation coefficient does *not* check for lurking variables, guard against extrapolation or against the ecological fallacy, or tell whether x causes y.

Cautions and warnings: A good fitted relationship with high correlation makes it tempting to jump to conclusions. Here are four things to look for before you leap.

1. *Extrapolation is risky.* Don't assume that an observed relationship holds outside the range of your data.
2. *Lurking variables.* The x, y relationship you see may involve some unobserved third variable.
3. *The "ecological" fallacy.* A relationship that holds for group averages may not hold for individuals.
4. *Association is not causation.*

5.2 COMPARING GROUPS, OR ANALYSIS OF VARIANCE (ANOVA)

Response Is Quantitative, Predictor Is Categorical

Where does this topic fit in? Analysis of variance (ANOVA) deals with relationships between a categorical predictor and quantitative response variable: For each category, you have a group of response values, i.e., a distribution. The main **plots** are parallel dot plots and box plots (Topic 2). The ideal **shape** is for each group to be normal, with group spreads equal. For such plots, you can give numerical **summaries** by computing the mean response for each group (**fit**), and an overall SD (**deviation from fit**).

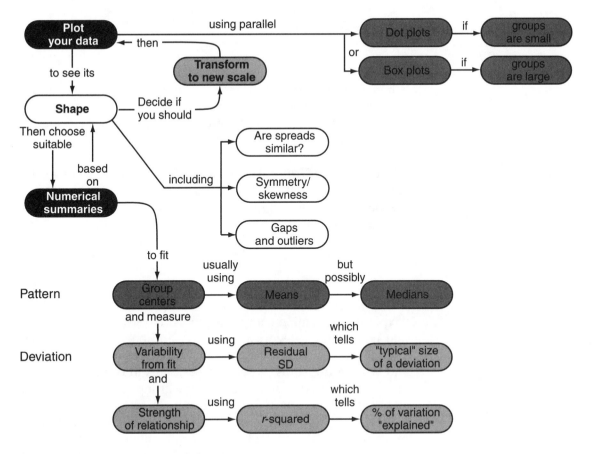

Describing Relationships: Quantitative Response, Categorical Predictor

Parallel Plots You can compare groups using the same plots you use to show distributions—dot, stem, and box plots. Put your plots in parallel columns or rows, one per group, using a single, common scale. For example, the video segment on the CD uses parallel box plots to compare caloric content (quantitative response) for three kinds (categorical predictor) of hot dog brands (cases).

Shape The ideal shape for a parallel plot has the following features. (The first three are the same as for a single distribution.)

1. *Modes.* The distribution in each group has only one mode.
2. *Symmetry/skewness.* The distribution in each group is symmetric.

3. *Tails, gaps, and outliers.* There are no long tails, large gaps, or outliers.
4. *Spreads.* The spreads of the distributions are roughly equal.

When spreads differ, it is less meaningful to compare groups using averages. However, if spreads are unequal and spread and average are related, then changing to a new scale for the response can make the spreads more nearly equal.

Fitted Values = Group Averages For each group (= value of the categorical predictor), the fitted value equals the average response value for that group. The residual is just the deviation from the group average.

$$
\begin{aligned}
\text{Obs} &= \text{Group Ave} & + & \quad \text{Deviation} \\
&= \text{Fit} & + & \quad \text{Res}
\end{aligned}
$$

Deviations: Measuring strength using fraction of variation "explained" Imagine comparing two analyses of the data: one that ignores the categorical predictor, and one that uses it. The second analysis, using the predictor, will give smaller residuals, and we can use sums of squared residuals to compare the two analyses. The first analysis ignores the group structure and puts all the data values into one big group. There is only one fitted value, the overall mean, and we can measure deviations from fit by computing the sum of squared deviations from the grand average. This sum of squares is the "total variation." The second analysis sorts cases into groups and computes a separate mean (= fit) for each group. The residuals from this analysis are the deviations from the group means; their sum of squares is the "residual variation." The change in the residual sum of squares (total variation minus residual variation) is the "variation explained" by the categorical predictor, and the fraction of variation "explained" (= [total − residual]/total) is mathematically equivalent to the r^2 you compute in a regression analysis. Values near 1 indicate a strong relationship; values near 0 indicate a weak relationship.

Caution If the fraction of variability accounted for by the group averages is large (a strong association), it can be tempting to jump to conclusions. Here, just as when you have a high correlation, caution is in order. The pattern of association may be due to some hidden cause. You should be especially careful when the groups are observational (already built in, rather than assigned using a chance device).

5.3 TWO-WAY TABLES OF COUNTS, OR CONTINGENCY TABLES

Response and Predictor Are Both Categorical

Where does this topic fit in? If both X and Y are categorical, you can display the data in a two-way table, with one row for each category of the X variable, and one column for each category of the Y variable. Numbers in the body of the table tell how many cases belong to each combination of row category and column category. Row totals, shown in the far right column, give the marginal counts for the X variable. Column totals, shown in the bottom row, give the marginal counts for the Y variable.

Plots for two-way tables Does the distribution of Y depend on X? Two kinds of plots—stacked bar charts and three-dimensional bar charts—allow you to compare the conditional distributions of Y for the various X categories. Stacked bar charts have side-by-side stacks of blocks, with one stack for each X category (row of the table). In each stack, there is one block for each Y category, with the height of the block equal to the count or percentage. Three-dimensional bar charts have the same layout as the table itself, with rows and columns for X and Y categories. The bar in each row and column has height equal to the count or percentage.

Pattern If both variables are categorical and the categories do not have a natural order, your analysis will be much more limited than in situations when at least one of your variables is quantitative. **Association** deals with the question, "If you want to predict the Y category, how useful is it to know the X category?" The **strength** of association is a feature of *any* two-way table. When an association is **weak**, X tells little about Y. The conditional distributions (in %) are nearly the same for all X categories, and so are nearly the same as the marginal percentages. When an association is **strong**, X tells a lot about Y. The conditional distributions (in %) differ a lot from one row to the next. (For the extreme case of perfect prediction, knowing the X category tells you the Y category.) For some tables, both X and Y have natural orderings, and the association may have a **direction**—positive or negative. If larger values of X are associated with larger values of Y, middle with middle, and smaller with smaller, the association is called **positive**. If larger values of X are associated with smaller values of Y, and smaller X with larger Y, the association is **negative**. A useful model that serves as a benchmark for measuring strength of association is the model of **no association**: X tells nothing about Y. More specifically, the true conditional percentages for Y don't depend on X, and so are equal to the marginal percentages for Y.

Deviations For the **model of no association**, each fitted value is equal to (# in row) × (% in column), which can be computed as (row total) × [(column total)/(grand total)]. For two-way tables, fitted values are often called **expected values**. We measure deviations from the model of no association in the usual way: Res = Obs − Fit, Dev = Observed − Expected. If the observed association between X and Y is weak, the model of no association will give a good fit, and residuals will all be near 0. If the observed association is strong, the model of no association will give a poor fit, and some, perhaps many, residuals will be far from 0.

Caution If the association between X and Y is strong—if knowing X is useful for predicting Y—it can be tempting to jump to conclusions, but be careful! Lurking variables may account for the pattern. For example, there is a moderately strong association between a state's region in the U.S. and its average verbal SAT score. The lurking variable? Percent of high school seniors who take the SAT. Even more striking, the pattern you see in a two-way table may hide an important relationship that is only apparent when you take a third variable into account.

5.4 LOGISTIC DATA

Response Is Categorical, Predictor Is Quantitative

Where does this topic fit in? There are four kinds of X, Y relationships, since each of X and Y can be either categorical or quantitative. Logistic regression deals with the combination of a categorical response and a quantitative predictor. Although this topic is not currently part of a typical first course, you can expect that to change over the next few years. There are two reasons. First, the topic is important. Studies with a quantitative predictor and categorical response occur frequently in medicine, in the law and public policy, and in basic research in all the sciences. Second, the topic has been avoided in beginning courses mainly because the calculations are too messy and long to do by hand. Now that computers are freeing us from the old notion that a first course shouldn't teach a statistical method unless you can do its calculations by hand, a major obstacle to teaching logistic regression has been eliminated.

Examples You can invent dozens of examples yourself of structures that call for logistic regression. Think of any of the yes/no variables you've studied in your course, and then imagine a quantitative variable that might reasonably be expected to influence the outcome. Were employees laid off or not? The chance of getting fired might well depend on an employee's age (even though age discrimination is illegal) because older employees tend to be paid more than younger ones. Did the patients die after surgery or not? The outcome might depend on the patient's age because younger patients tend to be in better physical condition. Did the savings and loans fail or not? The outcome might depend on the percentage of the bank's assets invested in real estate or on the bank's charge-off rate for the previous year. Did the citizens vote for or against the bond issue? The outcome might well depend on a person's family income.

Exploratory Analysis When you have a lot of data, you can plot y = % in a category against x = predictor. Often the **logistic transformation**, changing from y (= %) to $\log[y/(1 - y)]$ will give you a straighter plot.

5.5 BALLOON SUMMARIES: A VISUAL APPROACH TO REGRESSION AND CORRELATION

The Correlation Coefficient

Introduction. For scatterplots whose points form an oval balloon, it is possible to summarize the entire plot with numerical answers to just two questions. One: What is the equation of the fitted line that relates y-values to x-values? Two: How fat or skinny is the balloon? That is, how close do the points come to the line? We can answer the first question by giving the slope and intercept of the least squares line. An answer to the second question is given by the **correlation coefficient**. Its value is always between −1 and 1; its sign tells whether the balloon tilts up (+) or down (−); and its absolute value tells the shape of the balloon, with values near 0 for fat, directionless blobs, and values near 1 (+ or −) if all the points lie nearly on a line. A precise definition of the correlation coefficient will be given later, but first, here's a display of several scatterplots to give you a feel for what correlation measures.

Example 1. Scatterplots and Balloon Summaries for a Range of Correlation Coefficients

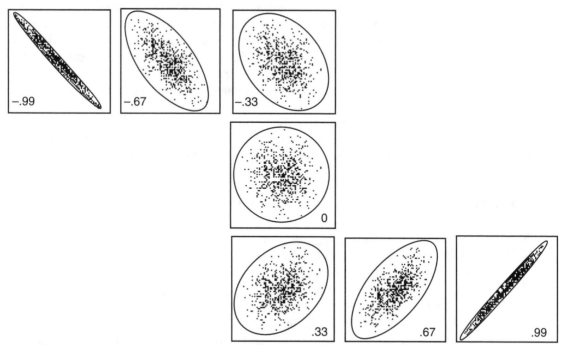

Correlation tells shape and direction for balloon summaries.
For balloon-shaped scatterplots, the correlation tells how well a regression line summarizes the x, y relationship. A correlation near 1 or −1 means the "balloon" is practically a line: regression gives very precise predictions (fitted y-values) for the response. A correlation near 0 means the balloon is almost round: the direction of the least squares line is not well determined by the data, and the line gives generally poor predictions.

The scatterplots in the figure have more than 600 points each. Often a data set will have far fewer, and the balloon shape of the plot will be less obvious. Nevertheless, it is reasonable to ask what the plot would look like if you had a lot more points. As you'll see in the next several paragraphs, one useful way to think about the correlation is to regard it as a description of the shape of the balloon that summarizes what you think the scatterplot would look like if you had many more points.

Like any one-number summary, the correlation has both advantages and disadvantages. A single number takes a lot less space (and attention) than a scatterplot, and so journal articles often report a correlation and don't show the scatterplot. For example, when psychologists report reliabilities, they almost never show the scatterplot.

One-number summaries also make comparison easier. For example, you can use four correlations to summarize the results of Carl Rogers' study of psychotherapy in Question 24. There, a higher correlation means that subjects saw themselves as more like the kind of person they wanted to be. The average correlation for subjects who got therapy went from .0 before therapy to .3 after. The average for subjects in the control group started out at .6, and didn't change. Before therapy, the way subjects in the treatment group saw themselves was unrelated to the way they wanted to be; after therapy, they saw themselves as somewhat closer to the way they wanted to be, but not as close as subjects in the control group.

Although the correlation is a compact summary that makes comparisons easier, it doesn't tell you about outliers, and what it does tell about the shape of the scatterplot is quite limited: it only tells how close to a line the points fall, and whether the line slopes up or down. If the points lie along a curve or in two distinct blobs, the correlation coefficient won't tell you that. There's another way to say this.

Correlations summarize balloons.
If your plot isn't balloon-shaped, don't use a correlation.

Balloon-Based Estimates for Scatterplots

For a "balloon-shaped" data set, you can find quick and quite accurate approximations to the least squares line and the correlation coefficient, working directly from a balloon summary. Although the geometric methods are approximate, they offer two advantages that make them worth learning. First, they are a lot quicker than using a calculator. Second, they can help you visualize what it is that the messier computing rules refer to. Moreover, in the idealized world of statistical theory, the geometric methods are exactly equivalent to the computing rules: one is as good as the other.

As an example for showing you how all this works, I'll use data that come from a study designed to compare the effects of heroin, morphine, and a placebo on the level of mental activity for human subjects, using only the data for the placebo. There are 24 cases, one for each subject. The response is the score on a 7-item questionnaire, given two hours after a placebo injection. (Scores range from 0 to 14; higher scores mean higher levels of activity.) The predictor variable is the mental activity level just before the injection, obtained from the same 7-item questionnaire as the response.

Drawing the Balloon Summary. To get the balloon summary, start with a scatterplot. Then draw a symmetric, oval balloon enclosing most or all of the points.

Step 1. Data = Scatterplot. The scatterplot in the figure below has far fewer points than the plots in Example 1, and the cluster of points shows some irregularities—the plot isn't symmetric; and the density of points doesn't show the same smooth pattern as those plots: densest in the middle, getting gradually less dense as you move away from the center.

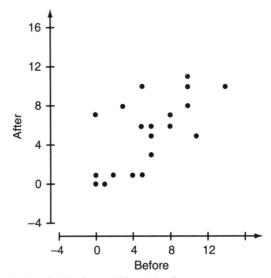

Scatterplot for the mental activity data.

What would the plot look like if we had much more data from the same source, in this case lots more subjects? Here are two possibilities.

1. If the irregularities are due to chance variation, then piling on lots more data would bury the first 24 points in a balloon-shaped cloud of hundreds more. According to this view, the balloon suggested by the 24 points is the message, and the irregularities are just noise.
2. On the other hand, the irregularities might be part of the message. It could be that the "true" relationship between Before and After readings under the placebo conditions is not as simple as a balloon summary suggests.

I hope you recognize the two possibilities, and the two ways of looking at the data, as variations on the old theme of Data = Fit + Residual. A good data analysis should try for a simple fit, but should look at the residuals, too. So keep in mind as you read the rest of this, that balloon summaries and correlations correspond to the fit and don't tell you about the residuals.

Step 2. Balloon = Fit. Draw a symmetric oval balloon, enclosing all or almost all the points. See the following figure.

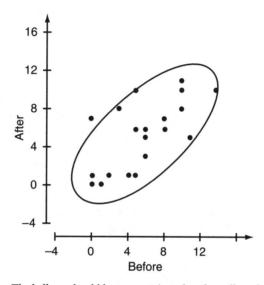

The balloon should be symmetric and enclose all or almost all the points.

Lines of Symmetry. Since our balloon summary assumes that the source of our data tends to produce a symmetric plot, we want our balloon to take advantage of that assumption. (If your first try at a balloon isn't symmetric, keep adjusting the shape until you get one that has two lines of symmetry.)

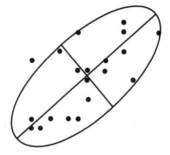

The balloon should have two lines of symmetry.
*The two lines of symmetry are the longest possible diameter, called the **major axis,** and the shortest possible diameter, called the **minor axis.** If you were to fold your balloon in half along either line of symmetry, the two halves should match.*

It might seem that the major axis ought to be the least squares line, but it isn't. To see why, and to find the least squares line, you need to think about vertical slices. To anticipate just a little, it turns out that how far the least squares line is from the line of symmetry depends on the shape of the balloon; comparing the two lines will give the first of three ways to measure shape.

The Correlation Coefficient and the Regression Effect

Interpretation: Vertical Slices. Turn back to Example 1, and imagine cutting a thin vertical slice through the scatterplot and balloon with correlation .67. What does the slice tell us? The slice corresponds to points whose *x*-values are about the same. The scatter in the vertical direction tells how the *y*-values are distributed when *x* is roughly constant. Notice two things about the pattern: The points are densest at the *midpoint* of the slice, and the density falls off in a symmetric way as you move away from the midpoint on either side. A histogram (Topic 2) would look roughly normal: mound-shaped and symmetric.

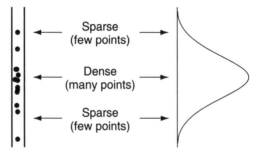

Distribution of y-values in a slice is roughly normal.

How can we use this information to predict y from x? Suppose you knew that a subject's mental activity level (x) before the placebo injection was 10, and you wanted to predict the level two hours after the injection (y). A reasonable predicted value would be the y-value for the midpoint of the vertical slice at $x = 10$. If the data behave like the scatterplots in Example 1, then the midpoint prediction corresponds to the y-value where the points are densest—the value under the peak of the normal curve.

It can be proved that this is in fact the least squares prediction. Any other choice would give a larger residual sum of squares. In particular, notice in the following figure that the major axis does not cut the vertical slices at their midpoints.

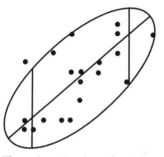

The major axis misses the midpoints of the vertical slices.
Slices to the right of the balloon's center have their midpoints below the line of symmetry.
Slices to the left have their midpoints above the line of symmetry.

The figure below shows how you can check that the set of midpoints for all possible vertical slices form a line. This line is the least squares line.

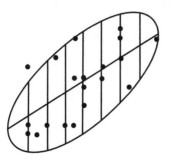

The line of midpoints is the least squares line.

A *Shortcut.* Notice that in the figure on page 119, the least squares line intersects the balloon at the two points where it is possible to draw vertical tangent lines. To see this, imagine taking vertical slices closer and closer to the right end of the balloon. Notice that the slices get shorter and shorter, shrinking to a single point at the end. The "slice" at that point corresponds to the vertical tangent. To find the line of midpoints, we only need two points on the line, so we can get the least squares line by finding the two points where there are vertical tangents and joining them with a line.

Step 3. *The Box of Tangents.* Draw the two vertical tangent lines and the two horizontal tangent lines, then connect them to form a box. Join the two points of vertical tangency to get the least squares line.

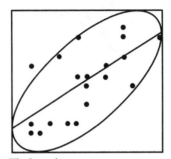

The box of tangents.
The least squares line joins the two points of vertical tangency

Step 4. *Least Squares Slope* = TAN$_y$MAX$_x$? The slope of any line equals "rise over run" or "change in y over change in x," so we can estimate the least squares slope by computing the change in y and the change in x between the two points of vertical tangency. See the figure on page 121.

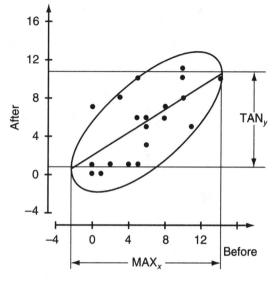

TAN$_y$ and MAX$_x$.

Let TAN$_y$ = rise = vertical distance between the two points of vertical tangency.

Let MAX$_x$ = run = horizontal distance between the same two points = width of the box of tangents. *Caution:* If your plot uses different scales for y and x, you'll need to measure TAN$_y$ and MAX$_x$ in those different scales. In this example, TAN$_y$ runs from about .75 to 10.75 along the y-axis, so TAN$_y$ ≈ 10. MAX$_x$ runs from about −2.25 up to 14.25, so MAX$_x$ ≈ 16.5.)

Slope = rise over run = TAN$_y$/MAX$_x$. In this example, the slope is about 10/16.5 ≈ .61. (The actual least squares slope is .6.)

Step 5. Correlation = TAN$_y$/MAX$_y$. How far apart will the least squares line and the line of symmetry be? Turn back to the plots of Example 1 and notice that the answer depends on how fat the balloon is. When the balloon is a perfect circle (correlation = 0), the two lines are as far apart as possible: the line joining the points of vertical tangency is horizontal, and the major line of symmetry is, as always, a diagonal of the box of tangents. At the other extreme, when the balloon is very skinny (correlation near 1 or −1), the two lines are almost on top of each other.

We can use this fact to measure the shape of a balloon summary. Refer to the figure on page 122. We express TAN$_y$ as a fraction of the length of the vertical side of the box. This gives us the correlation coefficient, apart from the sign. If the balloon is a perfect circle, TAN$_y$ will be 0, and so will the correlation. If the "balloon" is actually a line, then TAN$_y$ and MAX$_y$ will be equal, and their ratio will be 1. For the mental activity data, we get a correlation in between.

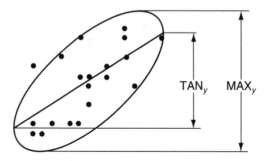

Correlation = TAN_y/MAX_y.
Since TAN$_y$ and MAX$_y$ are both measured vertically, they will always both be on the same scale, and you can measure them with a ruler. In this picture, TAN$_y$ = 25 mm, MAX$_y$ = 37 mm, and the correlation is 25/37 = .68.

Interpretation: The Regression Effect. What sort of predictions would we get if we used the major axis instead of the least squares line? Since we've defined correlation as a measure of how far apart (or close together) these two lines are, an answer to this question will give you another way to understand correlation.

We'll consider a particularly simple kind of situation, one where the x and y variables have equal averages and equal SDs. Then as you can see in the plots of Example 1, the line of symmetry is the 45° line whose equation is $y = x$. One example of a data set like the one just described was studied by the geneticist and statistician Francis Galton. His x-values were the heights of fathers, and his y-values were the heights of their adult sons. For Galton's data, the line of symmetry $y = x$ is the set of points with son's height = father's height. If you know that the father is 6'0", the prediction based on the line of symmetry is that the son will also be 6'0". Galton found that in reality, the average height for sons of six foot fathers was *less* than 6'0". On the other hand, the average height for sons whose fathers were only 5'4" turned out to be *greater* than 5'4". The pattern in Galton's data is an example of what we saw earlier: For vertical slices to the right of the center (fathers taller than the average), the midpoint of the slice (average height of their sons) was below the line of symmetry (less than their fathers' height). For slices to the left of the center (fathers shorter than average), the midpoint of the slice (average height of their sons) was above the line of symmetry (greater than their fathers' height). Galton recognized that on both sides of the average, the midpoint of a vertical slice would fall in between the line of symmetry and the overall mean for the population. Taller fathers tend to have sons who are not quite as tall. Shorter fathers tend to have sons who are not quite as short. Galton said the heights of the sons "regressed toward the mean." (And that's where the name "regression" for line fitting came from.)

Regression to the mean is a very general phenomenon, one that applies whenever a balloon summary is appropriate. It's well known in sports, for example. Teams (or individual players) with extremely good or extremely bad averages early in the season almost always end up drifting back toward the middle of the pack as the season wears on. It's also true of the mental activity data. On average, subjects with very high levels before the placebo injection have somewhat lower levels after; subjects with very low levels before tend to have somewhat higher levels after.

The correlation coefficient measures the size of the regression effect, as in the figure below. Large regression effects go with correlations near 0. Small regression effects go with correlations near 1 or −1. (Take a minute to verify this for the plots of Example 1.)

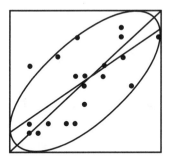

Correlation measures the size of the regression effect.
The least squares line has slope TAN_y/MAX_x, and the line of symmetry has slope MAX_y/MAX_x. The ratio of the two slopes equals TAN_y/MAX_y, the correlation coefficient.

Correlation and Fraction of Variability "Explained"

Overall and Conditional SDs for y. Our interpretation of the correlation coefficient in terms of the regression effect has been based mainly on comparing averages. There's a second interpretation, perhaps even more important than the first, based on comparing variability. This interpretation starts from two ideas you've already seen. First, a vertical slice through a scatterplot is basically just a dot plot for the *y*-values when *x* is fixed at a particular value. Second, the degree of spread in a dot plot is measured by the SD. Putting these two together lets us use the spread of a vertical slice to find the SD.

Extending these ideas will allow us to use the balloon for comparing two different kinds of SDs for the *y*-values. The first is the overall (or unconditional) SD for *y*, which I'll write SD_y, the one you get in the usual way, by regarding the whole set of *y*-values as a single sample. The other SD is the **conditional SD for *y* given *x***, written $SD_{y|x}$, which measures the spread of the *y*-values in the vertical slice at *x*. This SD takes the *y*-values in the slice as the sample, and you would compute this SD using the deviations from the midpoint of the slice.

Step 6. MID_y Tells the Conditional $SD_{y|x}$ for *y* When *x* Is Fixed. Let MID_y be the length of the vertical slice at the center of the balloon. MID_y tells the spread of the dot plot of *y*-values in the vertical slice at the midpoint, and is proportional to $SD_{y|x}$, the SD of the *y*-values when *x* is fixed. The more spread out the *y*-values, the bigger MID_y will be.

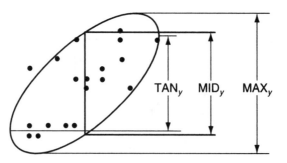

MID_y Tells the Conditional $SD_{y|x}$ of y given x.

A Caution. It's an obvious property of the balloon shape that vertical slices are longest at the center of the balloon, and get shorter as the slices get closer to either end of the balloon. Moreover, if you look at slices through one of the plots in Example 1, you find that dot plots for the slices extend about as far as the slices themselves. Dot plots for central slices extend farther than dot plots for slices near the ends of the balloon. It would be natural to think this means that the conditional $SD_{y|x}$ for *x* at the center is larger than $SD_{y|x}$ for *x* away from the center. Natural, but wrong.

> Same SDs
> A key assumption for least squares lines and balloon summaries is that $SD_{y|x}$ is the same for all values of x.

On the surface, it looks like the balloon summary and the assumption of same SDs are at odds, but they're not. Here's why it's possible for the assumption to be true and still have a balloon-shaped plot. Even though the dot plot for the central slice of the balloon is more spread out than the plots for slices away from the center, the extra spread results from having *more points* in the center slice. If just 1 of 20 points in a slice is more than 2 SDs from the average of the slice, then a slice with only 10 points might well have all of them within 2 SDs of the center; a slice with 50 points will probably have 2 or 3 points more than 2 SDs from the average, and so its dot plot will be more spread out even though it has the same SD.

If our assumption holds and $SD_{y|x}$ is the same for all x, we can use any vertical slice to measure $SD_{y|x}$. What's so special about MID_y? The answer is not obvious, but if you use MID_y to measure the conditional $SD_{y|x}$, then the ratio MID_y/MAX_y has a particularly useful interpretation.

Step 7. $(MID_y/MAX_y)^2$ = Fraction of Variation "Unexplained" by the Least Squares Fit. If MID_y measures the conditional $SD_{y|x}$, what can we use to measure to find the unconditional SD_y? The answer: MAX_y. I won't prove it, but I think you can convince yourself that it makes sense. Go back to the plots of Example 1, turn the page sideways, and pick one of the balloons. Now imagine that all the points in the scatterplot slide down and pile up along the y-axis to make a dot plot. The spread of this dot plot is what we measure with the unconditional SD_y, and this spread will be proportional to the overall vertical spread of the balloon, that is, to MAX_y.

Using a mathematical derivation based on these same ideas, you can show that the ratio of MID to MAX equals the ratio of the conditional SD to the unconditional SD.

$$\frac{MID_y}{MAX_y} = \frac{SD_{y|x}}{SD_y}$$

A useful interpretation comes from thinking about predicting y-values. If you don't know x, and you use the average of all the y's as your prediction, then SD_y is the SD for your prediction error, Obs − Pred = $y - y_{ave}$. If you know x, and you use the least squares fit (= midpoint of the vertical slice at x) as your prediction, then $SD_{y|x}$ is the SD for your prediction. If knowing x isn't much help, then $SD_{y|x}$ will be almost as large as SD_y, and the ratio will be near 1. On the other hand, if knowing x allows you to predict y with very little error, $SD_{y|x}$ will be small, and the ratio will be near 0. For technical reasons, it is more convenient to work with the square of the ratio.

$$\left(\frac{MID_y}{MAX_y}\right)^2 = \left(\frac{SD_{y|x}}{SD_y}\right)^2 = \text{fraction of variation not "explained" [1]}$$
$$\text{by the least squares line}$$

Notice that MID_y can never be larger than MAX_y, so the ratio MID_y/MAX_y will always be between 0 and 1. Values near 1 correspond to fat balloons—almost all the original variation in y remains "unexplained," still in the residuals after fitting the least squares line. Values near 0 correspond to very skinny balloons—only a tiny fraction of the original variation remains in the residuals after fitting the line.

We now have two ways to measure the shape of a balloon: the correlation (TAN_y/MAX_y) and the fraction of unexplained variation ($MID_y/MAX_y)^2$. It is reasonable to expect the two numbers to be related by a formula, since both measure different but related aspects of the shape of an ellipse. In fact, the two are closely related, and the formula relating them will give us the second interpretation of the correlation coefficient.

It can be proved mathematically that for any ellipse, $(TAN_y)^2$ and $(MID_y)^2$ add "like Pythagoras" to give $(MAX_y)^2$.

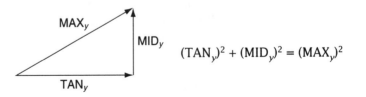

Dividing both sides of the equation by $(MAX_y)^2$ gives the following.

---

[1] **Fraction of variation "explained."** The quotation marks are meant to remind you not to assume that technical terms have their everyday meaning. For example, a least squares fit with x = temperature can "explain" a large fraction of the variation in y = frequency of cricket chirps, but that still doesn't explain, in the everyday sense, why faster rates go with higher temperatures.

$$\left(\frac{\text{TAN}_y}{\text{MAX}_y}\right)^2 + \left(\frac{\text{MID}_y}{\text{MAX}_y}\right)^2 = 1$$

Since $(\text{MID}_y/\text{MAX}_y)^2$ is the fraction *"un*explained," then $(\text{TAN}_y/\text{MAX}_y)^2$ must be (and is) the fraction of the original variation in y "explained" by the regression. The second interpretation of the correlation coefficient, then, is this.

(Correlation coefficient)^2 = Fraction of variation "explained"
by the least squares regression line

The Pythagorean relationship also gives a second way to estimate the correlation coefficient from a balloon summary.

$$\text{Correlation} = \pm \sqrt{1 - \left(\frac{\text{MID}_y}{\text{MAX}_y}\right)^2}$$

(Correlation)^2 = SS_{Regr}/S_{yy}. The balloon-based geometry has an exact parallel in terms of the sums of squares you get from decomposing each data value into pieces. You can think of the response y being split first into its average (y_{ave}) plus the "change in y" ($y - y_{\text{ave}}$); then the "change in y" is split into a fitted part plus residuals. The sums of squares for the three pieces add up. Here's the last two of the three pieces.

Obs:	change in y	=	(slope)(change in x)	+	residual
SS:	S_{yy}	=	SS_{Regr}	+	SS_{Res}

Decomposition of Response and SSs.
The "total variation in y," or sum of $(y - y_{\text{ave}})^2$, is split into a piece due to regression (SS_{Regr}) and a residual piece (SS_{Res}). The more nearly the points fall along a line, the bigger SS_{Regr} will be, and the smaller SS_{Res} will be.

We can express each of SS_{Regr} and SS_{Res} as a fraction of S_{yy}; the two fractions will then add to 1.

$$\frac{SS_{\text{Regr}}}{S_{yy}} + \frac{SS_{\text{Res}}}{S_{yy}} = 1 \quad (\text{because } S_{yy} = SS_{\text{Regr}} + SS_{\text{Res}})$$

The two fractions correspond exactly to $(TAN_y/MAX_y)^2$ and $(MID_y/MAX_y)^2$. Their two sizes tell how well or poorly a line fits the scatterplot:

Good Fit
Points near a line SS_{Regr} is a large proportion of S_{yy}, i.e.,
                              SS_{Regr}/S_{yy} is near 1.
                         SS_{Res} is a small proportion of S_{yy}, i.e.,
                              SS_{Res}/S_{yy} is near 0.

Poor Fit
Points in a fat blob SS_{Regr} is a small proportion of S_{yy}, i.e.,
                              SS_{Regr}/S_{yy} is near 0.
                         SS_{Res} is a large proportion of S_{yy}, i.e.,
                              SS_{Res}/S_{yy} is near 1.

SS_{Regr}/S_{yy} *measures how well a line fits the plot.*

Since $(SS_{Regr}/S_{yy}) = (TAN_y/MAX_y)^2$, we can translate the balloon-based rule for computing the correlation coefficient into a rule based on sums of squares.

$$\text{Correlation} = \pm\sqrt{SS_{Regr}/S_{yy}} = \pm\sqrt{1 - SS_{Res}/S_{yy}}$$

Example 2: Correlation as $\sqrt{1 - SS_{Res}/S_{yy}}$. For the mental activity data, we have the following.

$$
\begin{array}{ccccc}
S_{yy} & = & SS_{Regr} & + & SS_{Res} \\
266 & = & 118.8 & + & 147.2
\end{array}
$$

Thus the regression accounts for $118.8/266 = .45$ of S_{yy}, leaving $147.2/266 = .55$ of S_{yy} as residual SS. The correlation is $\sqrt{.45} \approx .67$.

Correlation Equals the Standardized Regression Slope

We now, as advertised, have two interpretations for the balloon-based correlation coefficient, one in terms of averages, and one in terms of SDs. There is one more important interpretation in terms of the slope of the least squares line. To set the stage, remember the warning about finding the least squares slope from a balloon summary. If x and y are measured in different scales, you need to use the x-scale to measure (change in x) = MAX_x, and use the

y-scale to measure (change in y) = TAN_y. The fitted slope depends on the scales you use. (If you change your x units from feet to inches, all x-values get multiplied by 12, so MAX_x gets multiplied by 12, and Slope = TAN_y/MAX_x gets divided by 12.)

The correlation coefficient does *not* depend on the scale in this way. If you change your y units from feet to inches, all of TAN_y, MID_y, and MAX_y get multiplied by the same number (12), and the ratios MID_y/MAX_y and TAN_y/MAX_y don't change.

The final step in setting the stage is to notice that both the least squares slope and the correlation coefficient have the form $TAN_y/MAX_{something}$.

$$\hat{\beta} = \text{least squares slope} = \frac{TAN_y}{MAX_x} \qquad \frac{TAN_y}{MAX_y} = \text{correlation coefficient} = r$$

Switching from MAX_x in the denominator to MAX_y switches from the least squares slope to the correlation coefficient. Algebraically, the switch corresponds to multiplying by the ratio MAX_x/MAX_y (= SD_x/SD_y).

$$\hat{\beta}\left(\frac{SD_x}{SD_y}\right) = \left(\frac{TAN_y}{MAX_x}\right)\left(\frac{MAX_x}{MAX_y}\right) = \left(\frac{TAN_y}{MAX_x}\right) = r$$

Note that this gives a way to compute the correlation coefficient numerically. First compute SD_x, SD_y and $\hat{\beta}$; then find r.

$$\boxed{\text{Correlation coefficient } r = \hat{\beta}(SD_x/SD_y).}$$

To get the third interpretation, notice that if $SD_x = SD_y$, then their ratio equals 1, and the least squares slope equals the correlation coefficient.

Ordinarily, of course, (x, y) pairs don't come to us with $SD_x = SD_y$. But if we compute SD_x and SD_y, then divide each x by SD_x and each y by SD_y, the new, rescaled x's and y's will have $SD_x = SD_y = 1$, and so the correlation coefficient r will equal the least squares slope $\hat{\beta}$. Usually, if you're going to divide by the SD you first subtract the average, then divide by the SD, to get a new **standardized** variable whose average is 0 and SD is 1. This is the basis of the third interpretation of the correlation coefficient, which also gives a second computing rule.

First standardize x and y. Replace each x-value by $\dfrac{x - x_{\text{ave}}}{SD_x}$ and each y-value by $\dfrac{y - y_{\text{ave}}}{SD_y}$.

The least squares slope for the standardized x and y equals the correlation coefficient for x and y.

You can see this relationship between the shape of a balloon and its least squares slope in the plots of Example 1. For these plots, although no scales are shown, the x- and y-values have already been standardized, so that $SD_x = SD_y$, and thus the least squares slope equals the correlation coefficient. Check, for example, that for the first plot in the last row, with correlation of $\frac{1}{3}$, $TAN_y = \frac{1}{3}MAX_y$, and since $MAX_y = MAX_x$, the slope $TAN_y/MAX_x = \frac{1}{3}$.

Example 3: Correlation = Standardized Regression Slope.
The standardization is easier to see with artificial data.

						Ave	SD
x	-3	-1	1	1	2	0	2
y	-5	-2	-1	5	3	0	4

For these numbers, $SD_x = \sqrt{16/4} = 2$, $SD_y = \sqrt{64/4} = 4$, $S_{xy} = 27$, and $\hat{\beta} = 27/16 \approx 5/3$. The y-values are twice as spread out as the x-values, which makes the least squares slope twice as big as it would be if the SDs were equal.

To find the correlation, we first divide each x by SD_x $(= 2)$, each y by SD_y $(= 4)$.

						Ave	SD
x	$-\frac{3}{2}$	$-\frac{1}{2}$	$\frac{1}{2}$	$\frac{1}{2}$	$\frac{2}{2}$	0	1
y	$-\frac{5}{4}$	$-\frac{2}{4}$	$-\frac{1}{4}$	$\frac{5}{4}$	$\frac{3}{4}$	0	1

For these standardized numbers, $SD_x = \sqrt{1} = 1$, $SD_y = \sqrt{1} = 1$, $S_{xy} = \frac{27}{32}$, and $\hat{\beta} = \frac{27}{32} \approx \frac{5}{6}$. The least squares slope $(\frac{27}{32} \approx .83 \approx \frac{5}{6})$ is the correlation coefficient. See the figure on page 131.

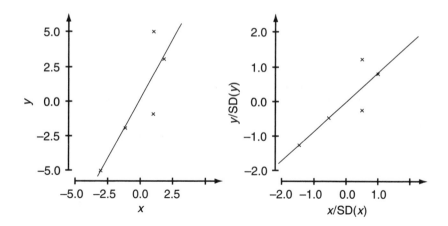

Correlation coefficient equals standardized regression slope.
For the plot on the left, in the original scale, $SD_y = 4$, $SD_x = 2$, and the least squares slope is
$5/3 \approx 1.67$. The plot on the right shows y/SD_y versus x/SD_x. For these standardized data, the
SDs are equal and the least squares slope is $5/6 \approx .83$, the correlation coefficient. For the left-
hand plot, the ratio $SD_y/SD_x = 2$, and the slope is twice what it would be if the SDs were
equal. The correlation coefficient equals (SD_y/SD_x) times the least squares slope.

A Final Caution: Three Traps to Avoid. Because the correlation coefficient is so useful as a summary, it can be easy to forget that, like the least squares line, it has limitations. Like regression, the correlation is designed for data pairs whose scatterplot suggests a line or balloon. *The value of a correlation doesn't tell you anything about whether the plot is actually balloon-shaped, and if it isn't, the correlation can be misleading as a summary.* The only insurance is to look at the plot.

Correlation, like regression, helps you study relationships among numerical variables, but *association is not causation.* In particular, a high correlation may be due to a lurking variable. For example, the correlation between the number of deaths in a state and the state's total tax revenue is .95. The explanation is not that taxes are life-threatening, but that both variables are roughly proportional to a state's total population. If you divide by population size, to get death *rates* and tax revenue *per capita*, the correlation gets much smaller, and even changes sign , to −.38.

This example reminds us that correlation is *not* causation, and teaches a second lesson as well: *High correlations don't always tell you more than low correlations.* The high positive correlation between deaths and taxes has no value as information because both variables are basically disguised versions of the same one variable—the state's population. The much lower correlation between death rate and per capita tax revenue may or may not be useful, but at least it does provide information that wasn't obvious beforehand.

Self-Testing Questions

1. The numbers below come from a study of lung cancer patients. Nine patients who survived nineteen months or less were compared with nine others who survived longer than nineteen months. The numbers give the morphometric index, a characteristic of tumor tissue.

Short term: 123   114   99   103   83   106   113   109   95
Long Term: 142   117   106   170   116   148   226   190   165

   (a) Draw a parallel dot graph, and use it to discuss the data.
   (b) Use the balance point interpretation to estimate the average for each group; then guess/estimate (do not compute) the SD separately for each group, choosing from 1, 5, 10, 20, 40, 60, and 100. What do your estimates tell you?

➤ *Solution*

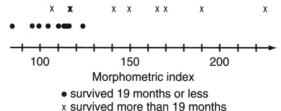

   • survived 19 months or less
   x survived more than 19 months

(a) The morphometric index is much lower (ave ≈ 100) for the short-term patients than for the longer-term patients (ave ≈ 150) suggesting that the index might be useful for predicting survival. Although there are two "straggling" values (195 and 226) for the long-term group, they are not so far from the rest of the group that they should be judged outliers, particularly in view of the patterns that suggest transforming. The SDs for the two groups are quite unequal (roughly 10 for short term, 40 for long term), with larger SD and larger average going together.

(b)

Survival	Average	SD
≤ 19 months	105	11.8
> 19 months	153.3	39.0

2. The data set below comes from an honors project done at Mount Holyoke by Kelly Acampora. The numbers record the specific activity of an enzyme used for transmitting nerve impulses. Four hamsters were raised on a cycle of 16 hours of light, 8 hours of darkness ("Long days"); the other four ("Short days") got the reverse, to simulate the approach of winter and time for hibernation.

	Hamster	Heart	Brain
Long Days	#4	246	394
	#1	169	297
	#7	186	216
	#2	183	373
Short Days	#3	468	216
	#8	390	198
	#5	314	194
	#6	508	192

Draw a parallel dot plot for the four combinations of day length and organ.

Answer

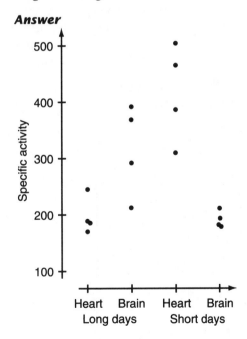

3. Compute the average and range, using the same four groups, and comment on what you find. Are the four ranges roughly equal? Does the range seem to be related to the average by a pattern?

Answer

Day length	Organ	Average	Range
Long	Heart	196	77
	Brain	320	178
Short	Heart	420	194
	Brain	200	24

The ranges are quite unequal. Larger ranges go with larger averages, and vice-versa.

4. Draw a scatterplot for the specific activity data plotting heart (y) versus brain (x).
(a) Are there any obvious outliers?
(b) Which of the following best describes the plot?
   (i) All eight points lie near a single line.
   (ii) All eight points form a single balloon-shaped cloud.
   (iii) The points form two distinct clusters, one for long days, one for short.
   (iv) None of the above.

Answer
(a)

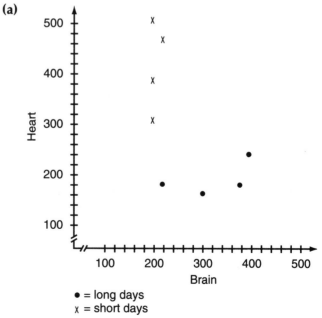

● = long days
x = short days

(b) (iii) The points form two clusters.

5. One of the main reasons for transforming your data to a new scale is to make the assumption of same SDs more appropriate.

    The specific activity (Problem 2) data had observations made under four conditions.

Long Days	Heart: 246, 169, 186,183	Brain: 394, 297, 216, 373
Short Days	Heart: 468, 390, 314, 508	Brain: 216, 198, 194, 192

(a) Look over the numbers and estimate (do not compute).

    (i) Which group has the largest average? Next largest? Smallest?

    (ii) Which two groups have the largest SDs? Which group has the smallest SD?

    (iii) Do larger SDs tend to go with larger averages, i.e., does the error size seem to depend on the size of the measurement?

(b) Compute the average and SD separately for each group. Are average and SD related? Should this data set be transformed? Why, or why not?

Answer

Here are the actual averages and SDs:

Day Length	Organ	Average	SD
Long	Heart	196	34
	Brain	320	81
Short	Heart	420	86
	Brain	200	11

The pattern here is the same as in Question 3: spreads are very unequal, with average and spread related. The data should be transformed.

    The table on p. 136 gives values of 10000/(specific activity). (numbers marked * were changed by ± 1 or ± 2 to simplify the arithmetic.) Notice that the change of scale has reversed directions: large values of the specific activity correspond to small values of 10000/(specific activity), and vice-versa.

OBSERVED VALUES

Heart	Brain	Hamster Averages
41	25	33
58*	36*	47
54	46	50
55	29*	42
20*	46	33
24*	50	37
32	52	42
20	52	36

CONDITION AVERAGES

	Heart	Brain	Day Length Averages
Long	52	34	43
Short	24	50	37
Organ Averages	38	42	40

Grand Average: 40

A Complete Set of Averages for the Transformed Hamster Data

6. Construct a parallel dot plot and compare it with the one from Question 2. Are the spreads more nearly equal now?

Answer

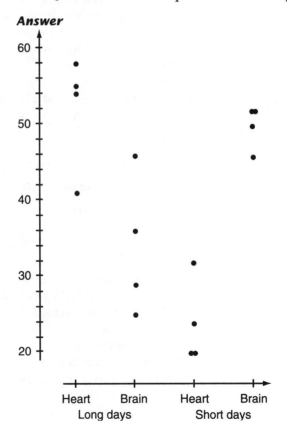

Spreads are now much more nearly the same.

7. Scatterplot brain values versus heart values, using separate symbols for long and short days. Does your plot show positive or negative relationships?

Answer

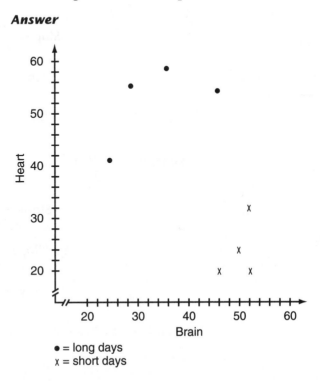

● = long days
x = short days

For all eight points taken together, the relationship is negative. Within each group of four, the pattern suggests a positive relationship.

8. *Fitting lines by eye: simple data for quick practice.* Four artificial data sets (a)–(d) are shown below. For each one, plot the pairs (x, y) and fit a line by eye. Then give the coordinates of two points on the line. Next, use the coordinates to compute the slope of the line. Finally, write the equation of the line in the point-slope form and in the form y = (intercept) + (slope)(x).
   (a) (x, y): $(0, 1)$, $(0, 3)$, $(4, 1)$
   (b) (x, y): $(0, 2)$, $(4, 2)$, $(4, 4)$
   (c) (x, y): $(0, 2)$, $(0, 4)$, $(3, 0)$
   (d) (x, y): $(0, 3)$, $(0, 4)$, $(2, 0)$, $(3, 1)$, $(3, 2)$, $(4, 2)$

Answer

(a)

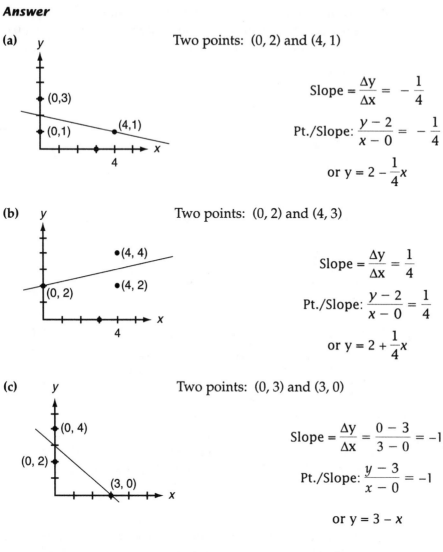

Two points: $(0, 2)$ and $(4, 1)$

$$\text{Slope} = \frac{\Delta y}{\Delta x} = -\frac{1}{4}$$

$$\text{Pt./Slope: } \frac{y - 2}{x - 0} = -\frac{1}{4}$$

$$\text{or } y = 2 - \frac{1}{4}x$$

(b)

Two points: $(0, 2)$ and $(4, 3)$

$$\text{Slope} = \frac{\Delta y}{\Delta x} = \frac{1}{4}$$

$$\text{Pt./Slope: } \frac{y - 2}{x - 0} = \frac{1}{4}$$

$$\text{or } y = 2 + \frac{1}{4}x$$

(c)

Two points: $(0, 3)$ and $(3, 0)$

$$\text{Slope} = \frac{\Delta y}{\Delta x} = \frac{0 - 3}{3 - 0} = -1$$

$$\text{Pt./Slope: } \frac{y - 3}{x - 0} = -1$$

$$\text{or } y = 3 - x$$

(d)

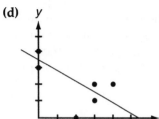

There is no obvious best choice for a line fitted by eye. Here's one:

Two points: $(0, 3\frac{1}{2})$ and $(3, 1\frac{1}{2})$

$$\text{Slope} = \frac{\Delta y}{\Delta x} = \frac{1\frac{1}{2} - 3\frac{1}{2}}{3 - 0} = -\frac{2}{3}$$

$$\text{Pt./Slope: } \frac{y - 3\frac{1}{2}}{x - 0} = -\frac{2}{3}$$

$$\text{or } y = 3\frac{1}{2} - \frac{2}{3}x$$

9. *Smoking and hearts.* The statistical evidence that smoking increases your risk of heart disease takes various forms. As a general rule, the evidence that is easier to get is not as compelling as other evidence that takes a lot more time and effort to collect. This problem is based on evidence of the quick-and-easy kind.

The points in the scatterplot below represent 21 countries. The *x*-coordinate gives the country's average cigarette consumption per adult per year, and the *y*-coordinate gives the country's mortality rate from coronary heart disease in deaths per 100,000 for adults age 35–64. The data are from the early 1960s, before cigarette packs and advertisements were required to display a warning from the Surgeon General.

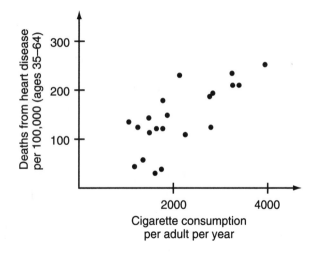

(a) Fit a line by eye to the scatterplot, and give the *y*-coordinates of the points on your line whose *x*-coordinates are 0 and 4000.

(b) Use the two points from part (a) to compute the slope of your line. Be sure to give the units of measurement.

(c) According to your fitted line, what would be the death rate for a country in which no one smoked? For a country with average consumption of 2000 cigarettes per person per year? (2000 is about 2 packs per person per week, but the average includes non-smokers as well as smokers.)

(d) An increase of 1000 cigarettes per person per year corresponds to an increase of how many deaths per 100,000?

(e) People smoke; countries don't. But the points of this plots correspond to countries, not people. In each country, some people smoke a lot, some smoke a little, some not at all. Would you find the data more persuasive if each point corresponded to a group of people who all smoked the number of cigarettes given by the *x*-coordinate of the point? Why, or why not?

Answer

(a) Two pts. (0, 0) and (4000, 250)

(b) Slope = $\frac{250}{4000}$ = $\frac{1}{16}$ ≈ .06 deaths per 100,000 people for each cigarette per person per year
More meaningful units: 20 cigarettes = 1 pack
$\frac{250}{4000}$ = $\frac{25}{40}$ = 12.5 deaths per 100,000 people for each pack per adult per year

(c) 0 deaths per 100,000
125 deaths per 100,000

(d) About 62.5 deaths per 100,000

(e) The data would be more persuasive if there was less variability within each case (= group of people).

10. *Cabbage Butterflies.* The time it takes a newly hatched butterfly larva to reach pupa stage depends on the surrounding temperature. It seems natural to measure time in days, and to record, for example, that at 20.5°C it took one larva 20 days to reach pupa stage. This scale for time scale, though natural-seeming, gives a curved plot. If, however, we change from # days to 1/(# days), and record that at 20.5°C, one day is $\frac{1}{20}$ of time it took to reach pupa stage, the data points form a straight line suitable for regression analysis.

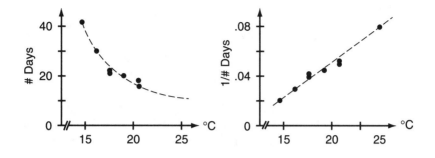

Fit a line by eye to the right-hand scatterplot above.

(a) Use your line to find fitted values for $y = (1/\text{\#days})$ for $x = 10°C$ and $x = 25°C$.

(b) Use the two points from part (a) to compute change in y, change in x, and the slope of the line.

(c) Write the equation of the line in point-slope form, with $y = 1/(\text{\#days})$ and $x = °C$.

(d) The time it takes, from egg hatch to pupa, is described by an equation of the form

$$(\text{\#days})(\text{temp}°C- \text{threshold}) = \text{constant},$$

where "threshold" and "constant" are numbers that depend on the species of butterfly. Use your fitted line to find the values of these two numbers for the cabbage butterfly. (Start with the equation of the line in (c), put $(1/\text{\#days})$ in place of y, and then rewrite the equation in a form that puts both $°C$ and # days on the left-hand side.)

Answer

(a) $x = 10° \Rightarrow y = \left(\dfrac{1}{\text{\# days}}\right) = 0$

$x = 25° \Rightarrow y \approx .08+$, or # days ≈ 12

(b) slope $= \Delta y/\Delta x\ x\dfrac{12 - 0}{25 - 10} \approx 0.80$

(c) $y - 0/x - 10 = .8 \Rightarrow y \approx .8x - 8$

(d) $\dfrac{1}{\# \text{ days}} \approx .8 \,°C - 8 \Rightarrow (.8 \,°C - 8)(\# \text{ days}) = 1$

or $(\# \text{ days})(\,°C - 10) = 1.25$

11. *Sleeping shrews: the effect of an influential point.* Scientists who study sleep have identified three kinds of sleep that can be distinguished by the kind of brain waves that occur: light slow-wave sleep (LSWS), deep slow-wave sleep (DSWS), and rapid eye movement sleep (REMS). Research on humans has established that our dreaming occurs during REM sleep.

For this study, the three kinds of sleep were the conditions of interest, and the subjects were six tree shrews. The response was heart rate, in beats per minute.

	CONDITION		
Block (shrew)	**LSWS**	**DSWS**	**REMS**
I	14.0*	11.7	15.7
II	25.8*	21.1	21.5
III	20.8*	19.7	18.3
IV	19.0	18.2	17.1*
V	26.0*	23.2	22.5
VI	20.4*	20.7	18.9

Heart Rate (Beats per Minute) of Sleeping Shrews

*Observations marked with an asterisk were changed by ± .1 to simplify arithmetic.

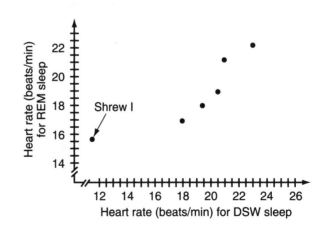

The scatterplot above shows heart rates for REM versus DSW sleep. For Shrew I, the heart rate for deep slow-wave sleep (the *x*-value) is far from the average, which means that the point has a lot of influence on the estimated slope.

(a) Ignore the influential point and fit a line by eye to the other five points. Find the coordinates of two points on your line, and use them to estimate the slope of the line.

(b) Now start over again, and this time fit a line by eye to all six points. Find the slope of this new line.

(c) Imagine fitting yet another line to the points, this time leaving out a point other than the outlier. Would the slope of this line be much different from the one in part (b)?

Answer

(a) Without the point for Shrew I, the line through the two extreme points (18.2, 17.1) for Shrew IV, and (23.2, 22.5) for Shrew V is one reasonable choice, which gives a fitted slope of about 0.9:

$$\text{slope} = \frac{\text{change in } y}{\text{change in } x} = \frac{22.5 - 17.1}{23.2 - 18.2} = \frac{4.4}{5.0} = 0.88.$$

(b) With Shrew I included, the fitted line will be quite a bit less steep. For example, if you choose the line through (12, 14) and (22, 22), your fitted slope is 0.8:

$$\frac{22 - 14}{22 - 12} = \frac{8}{10}$$

(c) No

12. *Chirping crickets.* Crickets have special organs on their front wings that make a chirping sound when they rub the wings against each other. As a rule, the warmer the air temperature, the faster they rub their wings. The relationship between temperature and chirp rate is well summarized by fitting a regression line, although the particular line depends on the species. Fifteen pairs of measurements for the striped ground cricket (*Nemobius fasciatus fasciatus*) give the following fitted line.

(# chirps per second) = 7.3 + (0.35)(Temp,°C)

(a) How many chirps per second would you predict at each of the following temperatures: 20°C, 25°C, 30°C?

(b) A temperature difference of 1°C corresponds to a difference (on average) of how many chirps per second?
What about a difference of 10°C? What about 10°F?

(c) Two of the data points were $(x, y) = (22, 16)$ and $(27, 16)$.
Find Res = Obs − Fit for each one.

(d) Find fitted values for the chirping rates at 0°C and
100°C. What do these numbers mean? What do they tell
you about the limitations of line fitting as a method for
data analysis?

(e) The value of S_{xx} = sum of (change in x)(change in x) for
this regression was 238.4. What was the value of sum of
(change in x)(change in y)?

(f) Rewrite the regression equation to correspond to the
following units.

	Temperature	Chirp rate
(i)	°C	Chirps/min
(ii)	°K (=°C − 273)	Chirps/sec
(iii)	°F	Chirps/sec

Answer

(a) 20° 14.3 per sec
25° 16.05 per sec
30° 17.8 per sec

(b) 1° corresponds to .35 chirps/sec.
10°C corresponds to 3.5 chirps/sec.
10°F corresponds to 3.5(5/9) ≈ 1.9 chirps/sec.

(c) $(x, y) = (22, 16)$. Fitted value = 15. Res = 1.
$(x, y) = (27, 16)$. Fit = 16.75, Res = −.75.

(d) At 0°C, the fitted line predicts 7.3 chirps/sec. At 100°C,
the fitted line predicts 42.3 chirps/sec. These fitted
values are unsound because crickets do not chirp at
these temperatures.

(e) $S_{xx} = 238.4$

Since slope $= \dfrac{S_{xy}}{S_{xx}} = 0.35$, $S_{xy} = .35$, $S_{xx} = 83.44$.

(f) chirps/sec = 7.3 + .35 (°C)
(i)   chirps/min = 438 + 21(°C)
(ii) chirps/sec = 7.3 + .35(°K + 273) = 102.85 + .35(°K)
(iii) chirps/sec = 7.3 + .35 (5/9 [°F − 32]) = 1.1 + .19(°F)

13. *Point of Averages = center of gravity.* Use the scatterplots (a)–(d) of Question 8 for this exercise.
  (i) For each one, guess the x and y averages, using the fact that, just as a dot plot balances at the average, a two-dimensional scatterplot balances at the point of averages, i.e., the coordinates of the center of gravity are (x_{ave}, y_{ave}).
  (ii) For each graph, write out the x and y coordinates of the points and compute the x and y averages to check your guess in (i).

Answer
(a) $(1\,\frac{1}{3}, 1\,\frac{2}{3})$
(b) $(2\,\frac{2}{3}, 2\,\frac{2}{3})$
(c) $(1, 2)$
(d) $(2, 2)$

14. *Fitting lines through the origin: simple numbers for practice.* Two data sets (a) and (b) are shown below. For each one, the point of averages is the origin $(0, 0)$, so the least squares slope is simply (weighted sum of y)/(weighted sum of x), using the x-values as weights. For each data set, plot the points and use your plot to estimate the slope of the least squares line. Then, compute the least squares slope and compare with your guess. (Some people's guesses tend to overestimate the least squares slope. Did yours?) Notice that set (b) is just set (a) with x and y switched. Why are the fitted lines not the same?

(a)

x	−2	−1	0	0	1	2
y	−1	−1	−1	1	1	1

(b)

x	−1	−1	−1	1	1	1
y	−2	−1	0	0	1	2

Answer

(a)

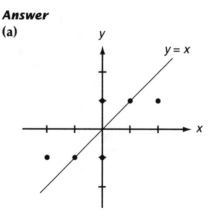

This is the line most people choose. It has slope = 1. The least squares slope is

(b) $$\text{Slope} = \frac{S_{xy}}{S_{xx}} = \frac{6}{10}$$

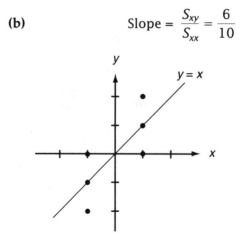

Here, too, people tend to choose the line y = x, which has slope =1. The least squares slope is

$$\frac{S_{xy}}{S_{xx}} = \frac{6}{6} = 1$$

15. *Estimates minimize SS_{Res}.* The purpose of this exercise is to illustrate an important general fact about the intercept and slope of a least squares line. If you were to use any other line, with a different intercept and/or slope, and compute

SS_{Res} for that line, you'd get a bigger number than the SS_{Res} for the least squares line. Use the data from Question 8 (c).

(a) Changing the intercept. Keep the slope fixed at the least squares value of –1, and compute residuals and SS_{Res} for lines with slope –1 and intercepts 1, 2, 2.5, 3 (least squares value), 3.5, 4, and 5. Notice that you can save time by finding the residuals geometrically from the scatterplot, as the vertical distances from the data points to the line. For example, the least squares line passes through (0, 3) and (3, 0), so the residuals are +1, –1, and 0.

(b) Draw a graph, with one point for each of your seven lines in part (a), plotting SS_{Res} on the y-axis versus the intercept on the x-axis. Join the points with a smooth curve, and use the curve to locate the intercept that minimizes SS_{Res}. Is this the same intercept as for the least squares line?

(c) Changing the slope. Keep the intercept fixed at 3, and compute residuals and SS_{Res} for lines with intercept 3 and slope equal to –2, –3/2, –1, $^{-1}\!/_2$, and 0. Here, as in part (a), you can save time by finding the residuals geometrically.

(d) Graph SS_{Res} versus slope, find the slope that minimizes SS_{Res}, and compare with the slope of the least squares line.

Answer

(a) y

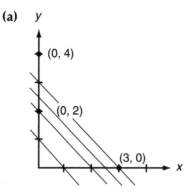

Intercept	SS_{Res}
1	$3^2 + 1^2 + (-2)^2 = 14$
2	$2^2 + 0^2 + (-1)^2 = 5$
2.5	$(3/2)^2 + (-1/2)^2 + (-1/2)^2 = 11/4$
3	$1^2 + (-1)^2 + 0^2 = 2$
3.5	$11/4$
4	5
5	14

(b)

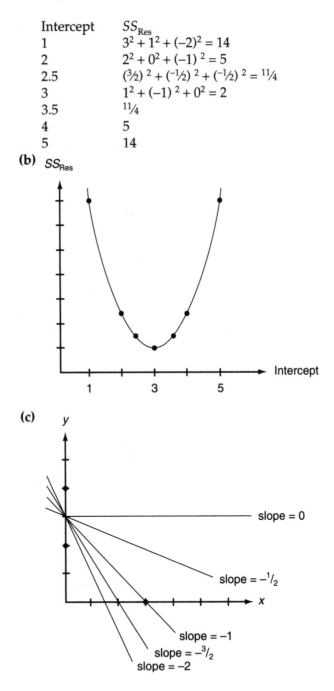

Slope	SS_{Res}
-2	$1^2 + (-1)^2 + (-3)^2 = 11$
$-\frac{3}{2}$	$1^2 + (-1)^2 + (-\frac{3}{2})^2 = 4\frac{1}{2}$
-1	$1^2 + (-1)^2 + 0^2 = 2$
$-\frac{1}{2}$	$1^2 + (-1)^2 + (\frac{3}{2})^2 = 4\,1/4$
0	$1^2 + (-1)^2 + 3^2 = 11$

(d)

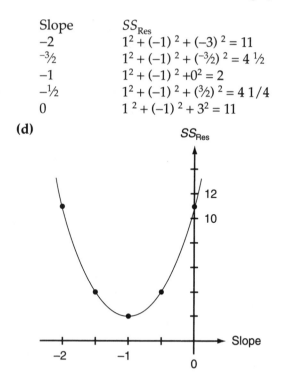

16. *The rise and fall of the portacaval shunt.* The portacaval shunt is an operation invented to help patients with bad livers. Shunting some of the usual blood flow to the liver through an artificial bypass decreases the workload on the sick liver. The operation was done so often that more than 50 papers were published reporting surgeons' evaluations of it.

Then, in 1966, the journal *Gastroenterology* published a study that summarized the earlier reports. Overall, they found that almost 70% of these earlier studies were markedly enthusiastic about the operation, another 20% were moderately enthusiastic, and only about 10% were unenthusiastic. On the surface, the operation looked pretty good. On the other hand, they also found that more than 60% of the studies had no control group, about 30% had a control group with serious flaws, and only four studies (fewer than 10% of the 51) had sound controls. Underneath the surface, the studies looked pretty bad. Things were actually even worse than that. For the 32 studies with no control group, enthusiasm was high: 24 of the 32 were markedly enthusiastic, only 1 was unenthusiastic. For the 15 studies with poor controls, the results were almost as positive: 10 of the 15 were markedly enthusiastic, only 2 were unenthusiastic. For the studies with good controls, however, the story was quite different: none showed marked enthusiasm, only one showed even moderate enthusiasm, and 3 (or 75% of the well-controlled studies) showed no enthusiasm. (After the article came out, the operation itself was quickly shunted to the dustbin of medical history.)

If you read carefully, you should be able to construct a table that gives the information from the article in *Gastoenterology,* the one that summarized the 51 previous studies. The purpose of your table should be to show the relationship between the soundness of a clinical study (the conditions) and the degree of enthusiasm for the operation (the response). Use as rows of your table the three kinds of controls (good, poor, and none); use as columns the degree of enthusiasm.

Answer

Controls	Marked	Moderate	None	Total
	DEGREE OF ENTHUSIASM			
Good	0	1	3	4
Poor	10	3	2	15
None	24	7	1	32
TOTAL	34	11	6	51

17. Twenty mothers of schizophrenic children were compared with twenty control mothers who had normal children of the same sex and approximately the same age. Each mother was shown the same set of ten pictures, and the ten stories she told were classified into categories by blind raters. The pictures were ones that were likely to lead to stories about parents. There were five categories, which had been developed in a similar but earlier study.

A. Personally involved, child-centered, flexible interactions
B. Impersonally involved, superficial interactions
C. Overinvolved, parent-centered interactions
D. One of A, B, or C, but can't tell which
IR. Irrelevant: none of A, B, and C applies

Here are the results:

Subj.	A	B	C	D	IR		Subj.	A	B	C	D	IR
1	2	2	4	1	1		1	8	0	1	0	1
2	1	0	2	1	6		2	4	0	0	1	5
3	1	1	4	1	3		3	6	0	1	0	3
4	3	1	1	1	4		4	3	0	1	1	5
5	2	1	0	1	6		5	1	0	2	1	6
6	7	0	0	0	3		6	4	0	2	1	3
7	2	1	2	2	3		7	4	1	2	1	2
8	1	0	1	0	7		8	6	0	0	0	4
9	3	2	0	1	4		9	4	0	1	0	5
10	1	1	3	0	5		10	2	1	2	0	5
11	0	1	4	0	5		11	2	1	2	1	4
12	2	1	3	0	4		12	1	1	3	2	3
13	4	1	1	0	4		13	1	1	3	0	5
14	2	2	1	0	5		14	4	0	1	1	4
15	3	0	3	1	3		15	3	0	2	1	4
16	3	0	1	0	6		16	3	0	2	0	5
17	0	0	3	1	6		17	2	1	0	2	5
18	1	1	3	0	5		18	6	0	0	0	4
19	2	0	4	0	4		19	3	1	1	0	5
20	2	1	2	2	3		20	4	0	2	0	4
Totals:	42	16	42	12	87		Totals:	71	7	28	12	82

MOTHERS OF SCHIZOPHRENICS (left) — **CONTROL MOTHERS** (right)

Construct two scatterplots, one for each set of twenty mothers. In the first scatterplot y = # type C stories; in the second, versus x = # type A stories, with each mother as a point; make two plots, one for each set of 20 mothers. Then write a short paragraph comparing the patterns in the two plots, and telling what the differences suggest about mothers of schizophrenics.

Answer

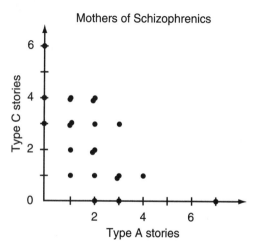

Mothers of Schizophrenics

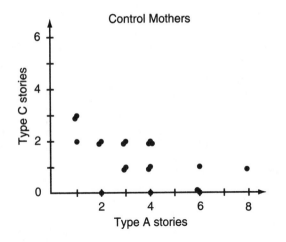

Both plots show negative relationships, which is not surprising, if you think about it: with ten stories total, the more of one type you tell, the fewer stories there will be left for the other types. The plot for the mothers of schizophrenics suggests a balloon, with a single outlier at (0, 7).

If you compare the two plots, you find that overall, mothers of schizophrenics tended to tell more Type C (overinvolved, parent-centered) stories than the mothers in the control group did. The mothers in the control group tended to tell more type A (child-centered, flexible) stories.

18. Match the following correlations with their scatterplots on page 153: .56, .03, −.90, .34.

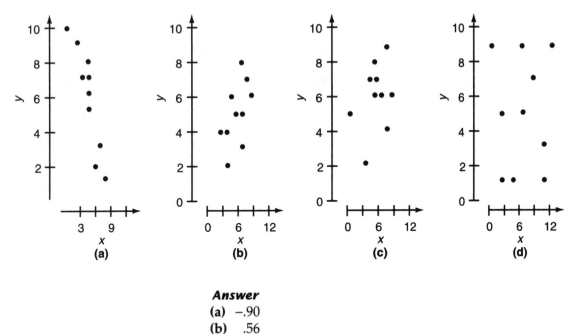

Answer
(a) −.90
(b) .56
(c) .34
(d) .03

19. *Mother's stories.* The scatterplots below are based on the data of Question 17. The plot on the left shows the numbers of Type C (overinvolved, parent-centered) versus Type A (personally involved, child-centered) stories told by mothers of schizophrenic sons. The plot on the right is for mothers in the control group.

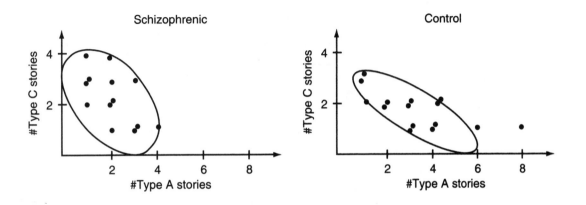

(a) Use the balloon summaries to match each plot as follows.

    (i) Give the slope of the least squares line, choosing from −.4 and −.6.

    (ii) Give the correlation coefficient, choosing from −.55 and −.85.

(b) Explain why the fact that each mother told a total of 10 stories all but guarantees the correlations would be negative.

(c) Try to find a plausible explanation for why one group of mothers shows a stronger relationship between numbers of stories than the other.

Answer

(a) **(i)** Least squares slopes: Control −.6, Schizophrenic −.4

    (ii) Correlations: Control ≈ −.85, Schizophrenic ≈ −.55

(b) The negative correlation is due to the fixed total, as described in the answer to part (a).

(c) Here's one possible explanation: The mothers in the control group are more consistent in the way they respond, which means they exhibit less variability in their total. That, in turn, leads to a stronger (x, y) relationship.

20. Estimate correlation coefficients for each of the following balloon diagrams. You'll need to measure with a ruler and apply the "balloon rule," don't just guess.

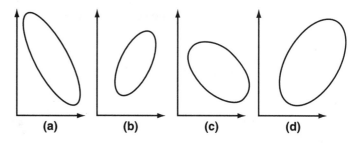

    (a)   **(b)**   **(c)**   **(d)**

Answer

(a) ≈ .79
(b) ≈ .54
(c) ≈ −.54
(d) ≈ .46

21. Would you expect the lengths and widths of a set of eggs to show positive, zero, or negative correlation? The scatterplot below shows length versus width for a dozen hens eggs.

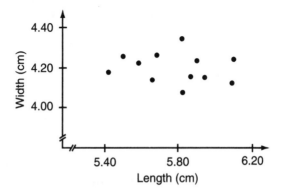

(a) Draw a balloon summary. You should use a pencil, and you'll probably have to make some changes to your first try in order to get a symmetric balloon. (Here are some things to check. Locate the center of your balloon. The vertical distance up, from center to the edge of the balloon, should be the same as the distance down, from center to edge. Similarly, the horizontal distance right, from center to edge, should equal the distance left.) Once you've got a symmetric balloon, use a ruler to find the following: TAN_y, MAX_y, and MID_y.

(b) Estimate the correlation two ways.
(i) As TAN$_y$/MAX$_y$,
(ii) As $1 - \text{RATIO}^2$, where RATIO = MID$_y$/MAX$_y$.
(c) Suppose that instead of a plot for 12 hens eggs, you had a plot for 12 eggs from 12 different birds, ranging in size from tiny to large. What sort of correlation would you expect? (Moral: The value of the correlation can depend a lot on whether your points correspond to cases that are similar or different.)

Answer
(a) TAN$_y$ = .12cm, MAX$_y$ = .34cm, MID$_y$ = .32cm
(b) (i) TÁN$_y$/MAX$_y$ $\approx$ $^{.12}/_{34}$ $\approx$ .35

    (ii) $\sqrt{1 - \left(\dfrac{\text{MID}}{\text{MAX}}\right)^2} = \sqrt{1 - \left(\dfrac{.32}{.34}\right)^2} \approx .34$

(c) With 12 different species, there would be a fairly strong, positive correlation.

22. *Smoking and hearts.* Draw a balloon summary for the scatterplot in Question 9. (Use the checks described in Question 21(a) to help make sure your balloon is symmetric.)
(a) Use a ruler to find TAN$_y$, MAX$_y$, MID$_y$, TAN$_x$, MAX$_x$, MID$_x$.
(b) Estimate the least squares slope as TAN$_y$/MAX$_x$, and compare with the slope of the line fit by eye in Question 9.
(c) Estimate the correlation coefficient two ways, just as in Question 21(b).
(d) Based on your experience drawing the balloon, which estimate do you have more confidence in? (Which of TAN$_y$ and MID$_y$ do you think is more reliable, given the way you draw balloons?)

Answer
(a) TAN$_y$ = 154, MID$_y$ = 169, MAX$_y$ = 231 deaths per 100,000
TAN$_x$ = 1,786, MID$_x$ = 2,071, MAX$_x$ = 2, 785 cigarettes/person/yr
(b) Least squares slope (est.) = TAN$_y$/MAX$_x$ = $^{154}/_{2,785}$ = .055 or 52 deaths/100,000 per 1,000 cig per person per yr
(c) Correlation: TAN$_y$/MAX$_y$ = .6%

$$\sqrt{1 - \left(\dfrac{\text{MID}_y}{\text{MAX}_y}\right)^2} = \sqrt{1 - \left(\dfrac{169}{231}\right)^2} = .68$$

(d) Neither seems obviously better than the other. If my two estimates don't agree, I redraw the balloon.

23. The data for Question 12 are shown in a scatterplot below, together with a balloon-summary of the plot. (Note that for this problem, the temperatures are in °F.)

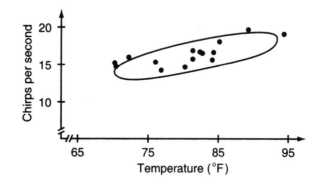

(a) Locate the center of the balloon, and use its coordinates to estimate the observed averages for x and y.

(b) Use the balloon method to find the approximated least squares line: Find the points of vertical tangency and use their coordinates to compute the slope of the line which passes through them. Then write the equation of the line in the form $(y - y_{ave}) = (slope)(x - x_{ave})$. (Is your slope roughly the same as the one you got in Question 12, part (f)iii?

(c) Use a ruler to measure TAN_y and MAX_y; then use these to estimate the correlation between chirp rate and temperature.

(d) Measure MID_y and use its value, together with the value of MAX_y, to estimate (i) SS_{Res}/S_{yy}, (ii) SS_{Res}/S_{yy}, and (iii) the correlation.

Answer

(a) $\bar{x} \approx 81.5°$, $\bar{y} \approx 16$ chirps/sec

(b) slope $= \Delta y/\Delta x \approx \dfrac{4.9 \text{ chirps/sec}}{23°F} = .21$ chirps/sec per °F

(chirps/sec $- 16) = .21$ (°F $- 81.5°$)

(c) $TAN_y = 1.0$cm, $MAX_y = 1.3$cm, Correlation $= {}^{2.1}\!\!/_{2.7} = .77$

Note: The balloon for the crickets is hard to measure well, so there are some discrepancies in your estimates.

(d) $MID_y = .8$cm

   (i) $SS_{Res}/S_{yy} \approx (MID_y/MAX_y)^2 \approx (\cdot^8/_{1.3})^2 \approx .38$

   (ii) $(SS_{Reg}/S_{yy}) = (TAN_y/MAX_y)^2 \approx .62$

   (iii) $Corr = \sqrt{1 - \left(\dfrac{MID}{MAX}\right)^2} = \sqrt{.69} \approx .83$

24. *Rogerian therapy and Q-sort.* The psychotherapist Carl Rogers used a method called the Q-sort procedure to compare people's assessments of "who I am" and "who I'd like to be," and used correlations to measure how closely the two assessments agreed. Here is a summary of his results.

	Start	End
Therapy group	0	.3
Control group	.6	.6

Average Correlations

For the therapy group, "Start" and "End" mean "before therapy" and "after therapy."

   Draw three balloon diagrams to represent scatterplots for an average subject in the therapy group, before therapy; the therapy group, after therapy; and the control group.

Answer

The correlation coefficient equals the ratio TAN_y/MAX_y, so you need to draw balloons with that ratio equal to 0, 0.3, and 0.6. Start by locating the points where your balloon will have its horizontal and vertical tangents. Then draw the balloon.

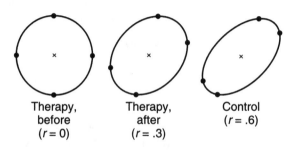

Therapy, before ($r = 0$)   Therapy, after ($r = .3$)   Control ($r = .6$)

What's wrong with this inference? The next several problems each describe evidence—a set of cases and two variables measured on those cases—and a conclusion based on the data. In each problem, the inference is not justified, sometimes because one or more lurking variables affects the correlation. For each problem, tell why the inference is not justified. Identify any lurking variable(s). Then tell what kind of study you would need in order to test whether the conclusion is in fact true.

25. Ham radios.

    Cases: Years, from 1920 to 1950

    Variables: Number of licensed amateur radio operators registered in Great Britain

        Number of officially certified mental defectives registered in Great Britain

    Conclusion: The high positive correlation means that amateur radio operators tend to be mentally defective.

Answer

Both x and y are roughly proportional to population size (the lurking variable).

26. Money and words.

    Cases: The twelve grades of a school district

    Variables: Average weekly allowance for all children in the grade

        Average vocabulary size for all children in the grade

    Conclusion: The high positive correlation means that raising children's allowances leads them to improve their vocabularies.

Answer

Both x and y increase with age of the kids (lurking variable).

27. Kids and blood pressure

Cases: A random sample of mothers, aged 20–45

Variables: Number of children born to the mother

Mother's blood pressure

Conclusion: The positive correlation shows that having kids raises your blood pressure.

Answer

Mother's age is a lurking variable: mothers with more kids tend to be older; blood pressure goes up with age.

28. Groceries and cancer

Cases: The 100 largest cities in the U.S.

Variables: Number of grocery stores in the city

Number of new cases of cancer in a year

Conclusion: The high positive correlation shows that the food sold at grocery stores tends to cause cancer.

Answer

Population of the city is a lurking variable.

29. Drunk driving

Cases: The 50 U.S. states

Variables: Annual sales ($) of beer, wine, and liquor

Number of arrests in a year for driving under the influence

Conclusion: Drinking alcohol leads to drunk driving.

Answer

Although the conclusion is true, the inference is not supported by the association between the two variables in this problem: state population is a lurking variable.

30. Videotapes and smoking

Cases: The years from 1980 to 1990

Variables: Percent of U.S. families owning a VCR

Percent of people over 18 who smoke cigarettes

Conclusion: The negative correlation shows that as more people were able to entertain themselves watching videotapes, they found it easier to give up smoking.

Answer

There are two time trends at work here: sales of VCRs went up during the 1980s, and cigarettes smoking declined, but there is no evidence of causation.

31. The purpose of this problem is to illustrate that the correlation equals the regression slope for x and y expressed in standard units. Find the correlation of x and y.

						Ave	SD
x	−3	−1	1	1	2	0	2
y	−5	−2	−1	5	3	0	4

➤ *Solution*

For these numbers, $SD_x = \sqrt{16/4} = 2$, $SD_y = \sqrt{64/4} = 4$, $S_{xy} = 27$, and $\hat{\beta} = {}^{27}/_{16} \approx {}^5/_3$. The *y* values are twice as spread out as the *x*-values, which makes the least squares slope twice as big as it would be if the SDs were equal.

To find the correlation, we first divide each *x* by SD_x (= 2), each *y* by SD_y (= 4).

						Ave	SD
x	$-^3/_2$	$-^1/_2$	$^1/_2$	$^1/_2$	$^2/_2$	0	1
y	$-^5/_4$	$-^2/_4$	$-^1/_4$	$^5/_4$	$^3/_4$	0	1

For these standardized numbers, $SD_x = \sqrt{1} = 1$, $SD_y = \sqrt{1} = 1$, $S_{xy} \approx {}^{27}/_{32}$, and $\hat{\beta} = {}^{27}/_{32} = 5/6$. The least squares slope $({}^{27}/_{32} \approx .83 \approx {}^5/_6)$ is the correlation coefficient.

32. The correlation between X, score on exam I, and Y, final exam score, in a college statistics course in the fall of 1987 was .8. The regression equation is

$$Y = 69.4 + 1.07X.$$

(a) Predict the final exam score of a student who scored 80 on exam I.

(b) Interpret, in the context of this setting, the value 1.07 from the regression equation. That is, what does the 1.07 mean?

(c) Interpret, in the context of this setting, the value 69.4 from the regression equation. That is, what does the 69.4 mean?

(d) In the context of this setting, interpret the value of r^2, which equals .64.

Answer

(a) The predicted score is $69.4 + 1.07 \times 80 = 69.4 + 85.6 = 155$.

(b) As score on exam I goes up by 1 point, final exam score goes up by 1.07 points, on average.

(c) For a hypothetical student who scored zero on exam I, the model predicts a score of 69.4 on the final exam.

(d) We can account for 64% of the variability between students in final exam score by using exam I score in a regression model.

33. (Based on the Berkeley longitudinal study.) Consider the two variables

X = height at age 9 (measured in centimeters)

and

Y = height at age 18 (measured in centimeters)

The regression equation is $Y = 45.6 + .89X$. The correlation between X and Y is .8.

(a) True or False. If height were converted from centimeters to inches the correlation would change.

(b) One subject, Linda, was 129 cm tall when she was 9 years old. What is her predicted height at age 18?

(c) Linda's actual height at age 18 was 157.1 cm. What is the residual for Linda?

Answer

(a) False. The correlation coefficient does not depend on the units of measurement used.

(b) In this case x = 129, so we predict y to be $45.6 + .89 \times 129$ $= 45.6 + 114.8 = 160.4$ cm.

(c) The residual is $160.4 - 157.1 = 2.7$ cm.

34. Consider the two variables

X = tuition in units of \$1000

and

Y = graduation rate,

where the data are from a sample of colleges. The least-squares regression line is given by $\hat{Y} = 38.5 + 2.7X$. The correlation between X and Y is .59. Thus, $r^2 = .35$

(a) Interpret the value 2.7 from the fitted model in the context of this problem.

(b) Consider a college with a tuition of \$18,000. Predict the graduation rate of this college.

(c) One college has a tuition of \$11,700. The residual for this college is 19.4. What is the graduation rate for the college?

(d) Interpret the value of r^2 in the context of this problem. That is, what does it mean to be told that $r^2 = .35$?

Answer
(a) As tuition goes up by 1 unit ($1000), graduation rate goes up by 2.7 (2.7%) on average.
(b) We have $x = 18$, so the prediction for y is $38.5 + 2.7 \times 18 = 38.5 + 48.6 = 87.1$.
(c) The predicted graduation rate is $38.5 + 2.7 \times 11.7 = 38.5 + 31.6 = 70.1$. The residual of 19.4 is the actual rate minus the predicted rate of 70.1. Thus, the actual graduation rate is $70.1 + 19.4 = 89.5$.
(d) We can account for 35% of the variability between colleges in graduation rate by using tuition in a regression model.

35. Here is a scatterplot of life expectancy and number of televisions per person for a sample of countries. The United States is the x toward the right-hand side of the plot.

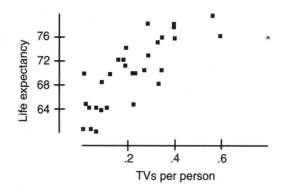

(a) Estimate the correlation coefficient.
(b) Suppose we were to fit a least-squares regression line to these points. Draw the residual plot that would result from the regression. Label the axes of your plot and identify the United States as an "x" on your plot.

Answer
(a) The correlation coefficient is .75.
(b) Here is a residual plot from the regression of life expectancy on TVs per person, with the United States shown as an x. The United States has a negative residual, which means that life expectancy in the United States is lower than the regression model predicts it to be.

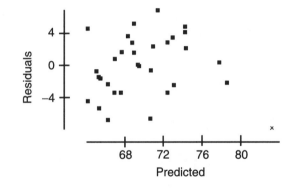

36. Recently an Oberlin College administrator pointed out that among high school students who have been admitted to Oberlin College, those who make a visit to campus are more likely to enroll than are those who never visit the campus. Identify the explanatory and response variables in this setting. Then explain what lurking variable is present and how this might prevent us from agreeing that "visiting campus causes students to be more likely to enroll."

Answer

The explanatory variable is "campus visit status" (whether or not someone visits the campus). The response variable is "enrollment status" (whether or not the person enrolls). The lurking variable is prior level of interest in Oberlin. Students who have a high level of interest in Oberlin are more likely to visit the campus, and more likely to enroll, than are those who have only a limited interest in the college. Thus, it could be that students who visit the campus and then enroll would have enrolled even if they had not visited the campus.

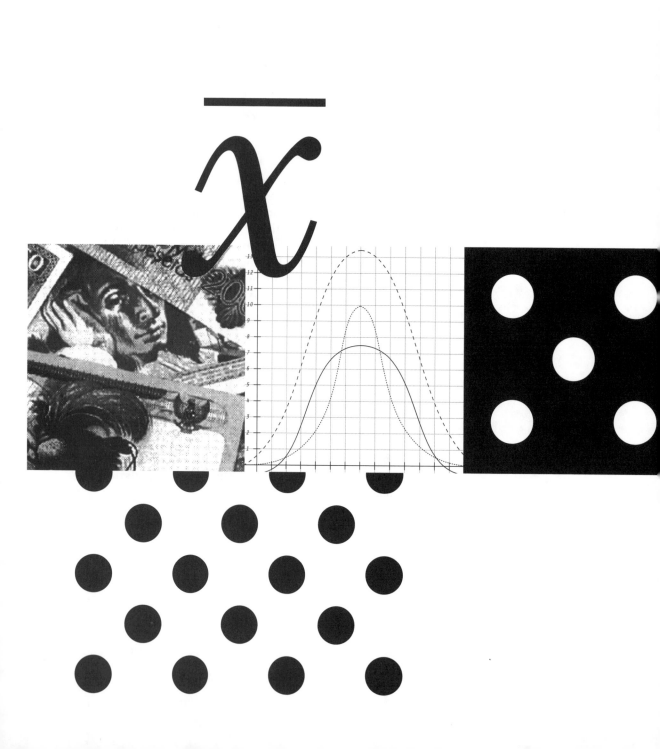

Producing Data

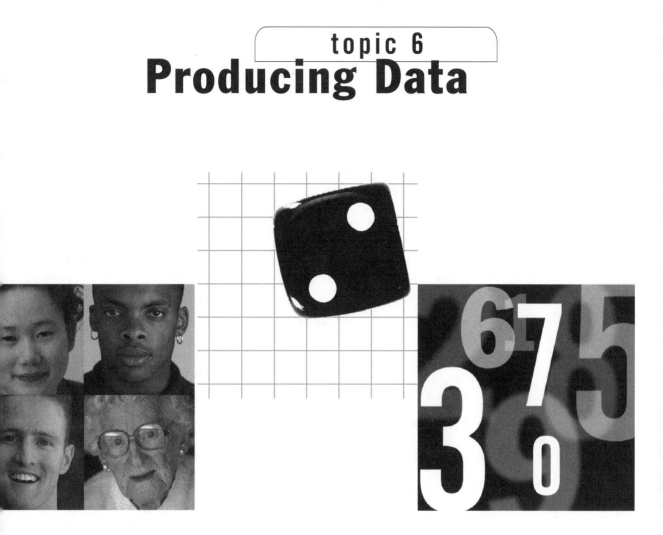

Summary

6.1 IN A NUTSHELL

Where does this topic fit in? Data analysis consists of two phases: **exploration** and **inference**. The bridge that joins these two is built of probability-based data **production** and the idea of imagining the results of a large number of **repetitions**. The first five topics have dealt with exploration and description—ways to uncover and summarize the important patterns in your data. Much of data analysis (and the final five topics of this workbook) deals with inference—ways to generalize about the patterns and use them to draw conclusions. This topic and the next three introduce the logic of inference. The basic idea is that the patterns in your data, and what they mean, depend in important ways on how your data were produced. Sloppy data production can easily give you misleading results. If you want to be able to draw sound conclusions, you must produce your data using probability-based methods.

What are probability-based methods for data production? We'll consider two sets of methods here: **random samples** and **randomized experiments**. For a random sample, you use a chance device to decide who gets into your sample and who doesn't. In a randomized experiment, you use a chance device to decide which treatment to assign to each subject.

Why do you get better data if you leave so much to chance? There are two main reasons; making sure you have a clear understanding of them both should be a major goal of your review.

1. Chance-based methods protect against **bias**.
2. Chance-based methods allow you to use probability theory to tell what would happen if you were to repeat the process (of producing and summarizing your data) a large number of times.

6.2 SAMPLES

Basic vocabulary of sampling The goal of scientific sampling is to allow you to generalize from the data you actually get to see—the **sample**—to some larger group—the **population**—that you would like to know about. (Here "population" can refer either to the set of individuals you want to know about, or to the set of values of a variable measured on those individuals.) In the same spirit, a **statistic** is a numerical summary you get to see because it is computed from the sample; a parameter is the corresponding numerical summary you would like to know about because it is computed from, and helps to summarize, the population. In real applications, there is usually a difference between the population, which is the target of your interest, and the list of individuals (the **frame**) you rely on to choose your sample.

Reasons for taking samples In principle, we'd like to look at the whole population, but in practice we usually can't. Sometimes the population is hypothetical, for example, the set of all possible rolls of two dice, or, for a similar but more practical example, the set of all outcomes of a manufacturing process. Sometimes, the measurement process destroys the objects we want to know about, as when you test the length of life of batteries. Sometimes, measuring every individual in the population (taking a **census**) is too expensive or too time-consuming. (Imagine trying to track down every single voter in the U.S. in order to predict the results of an election.)

Non-probability samples Samples that don't rely on probability can be tempting and may appear reasonable, but as a rule you should avoid them and be skeptical of results based on them. **Convenience samples**—taking whatever members of the population happen to be handy—are very common, but especially unreliable. In particular, **voluntary response samples** are notorious for their bias, because only those who care most about the issues go to the trouble of getting themselves into the sample. **Systematic samples**—for example, taking every fifth person from a list—can be biased if there is a systematic pattern in the list. **Judgment samples**—which rely on experts to choose a representative set of individuals from the population—can be biased in whatever ways the experts overlook or don't know about.

Probability samples A **probability sample** is one for which each individual in the population has a fixed chance, which you know in advance, of ending up in the sample. The great advantage of probability samples is that you can use either computer simulation or mathematical theory to repeat your sampling process dozens, hundreds, or even thousands of times, in order to study the properties of the sampling method, and the properties of any numerical summaries (statistics) you compute from the sample. (Take a minute to check that you can't use this idea of repetition to study the properties of judgment and convenience samples.) The simplest probability sample is the SRS, or **simple random sample**, for which each individual has the same chance as each other individual of ending up in the sample. A useful way to think about simple random samples is to imagine drawing tickets out of a box, with one ticket for each individual in the population. (In actual practice, it is very hard to mix the tickets well, and so it works much better to use random numbers, but the box of tickets is still a useful guide for your intuition.) There are two common and useful variations on the SRS: multi-stage samples, and stratified random samples. **Multistage samples** are useful when your population is so large and complex that it would not be practical to construct a complete frame to take a SRS. A three-stage sample of U.S. first-grade pupils might take a SRS of school districts in the whole country, and then an SRS of first-grade classes from each of the districts chosen, and then an SRS of kids from each of the classes chosen. **Stratified random samples** are useful when you want to make sure that all of several subgroups within your population are represented in the sample, and you are confident that differences from one subgroup to the next are larger than differences from one individual to another within single subgroups. For example, if you want to compare attitudes based on religious affiliation, an SRS would be unlikely to contain many people belonging to minority religious groups. If you wanted to make sure that Lutherans, Quakers, and Jews were well represented in a sample of North Carolina adults, you should use a stratified sample, taking several SRSs, one from each group of interest.

6.3 EXPERIMENTS

Basic vocabulary of experiments An experimental design is a plan for assigning **treatments**, the sets of conditions you want to compare, to **experimental units**, the subjects or chunks of material that receive the treatments and provide the values of the response variable. In a true experiment, you *assign* the treatments to units. If you can't assign them because the units come with the conditions of interest already built in (like smoker and non-smoker), you have an **observational study**, rather than an **experiment**. An experiment is always **comparative**—there is more than one set of conditions. For many experiments, one of the conditions is a **control**, a "non-treatment" you include to use as a baseline for judging the effect of the treatment(s) of interest. Often the control group receives a **placebo**—a non-treatment intended to keep subjects **blind** (unable to tell) about whether they are in a treatment or control group. In studies where measuring the response involves human judgment, the raters may also be kept blind regarding the treatments. Ordinarily, each treatment is **replicated**, assigned to more than just one unit, in order to be able to measure variability from one subject or unit to the next when conditions are otherwise the same. Finally, in a good experiment the assignment of treatments to units is **randomized** using a chance device. In brief, four features characterize a good experiment: comparison, assignment, replication, and randomization.

Randomization and the completely randomized design In a **CR (completely randomized) design**, you use a chance device to assign one treatment to each experimental unit. The design is **balanced** if each treatment is assigned to the same number of units. Randomization brings two main advantages: It protects against bias in the assignment of treatments, and it allows us to use computer simulation or probability theory to tell what would happen if we were to repeat the experiment a large number of times.

Blocking and the randomized complete block design If your experimental units can be put in groups of similar units—litter mates of lab animals, for example, or human subjects grouped by scores on a pre-test—you may be able to increase the efficiency of your experiment by taking advantage of these similarities. A **block** is a group of similar experimental units. A **randomized complete block (RCB) design** first sorts units into blocks, with the number of units in a block equal to the number of treatments. Then it randomly assigns one treatment to each unit in a block in such a way that each block of units gets a complete set of treatments, and the random assignment in one block is independent of the random assignments in the other blocks. The simplest RCB design is the **matched pairs design**: there are only two treatments to compare, units come in blocks of two (left and right eyes, twin pairs, etc.), and in each pair a coin toss or its equivalent decides which unit gets which treatment.

Factorial crossing, interaction, and the two-way factorial design The ideas here may be easiest to think about in terms of a tomato-growing example, which may not be terribly exciting, but has the virtue of being close to the roots of this part of the subject. (Sorry.) Suppose you want to compare the effect of nitrogen and phosphorus on the yield of tomato plants. You decide to compare three concentrations of nitrogen (low, normal, and high) and three of phosphorus (also low, normal, and high). In the language of design, nitrogen and phosphorus are **factors**, each with three **levels** (low, normal, high). In all, there are $3 \times 3 = 9$ combinations of levels of the two factors (low N low P, low N normal P, etc.).

The two factors are crossed if each of the possible combinations of levels gets assigned to some unit. A randomized, balanced **two-way factorial design** is one with two crossed factors, with each combination of factor levels randomly assigned to the same number of units. One of the main reasons for using factorial designs is to measure **interaction** between the factors. Two factors interact if the effect of one factor, as measured by the difference in response (say tomato yield for high nitrogen minus yield for low nitrogen), is different for different levels of the second factor (e.g., nitrogen has a larger effect when phosphorus is present at normal levels than when phosphorus is present at low levels.)

Self-Testing Questions

1. Fill in the blanks, choosing from bias and chance error.
   (a) A major reason for wanting several measurements under the same conditions is to be able to estimate the size of _____.
   (b) When _____ is present, all your measurements tend to be off in the same direction, either too high or too low.
   (c) _____ tends to go in different directions for different measurements, and to cancel when you compute averages.

   Answer
   (a) chance error
   (b) bias
   (c) chance error

2. True or false
   (a) An important reason why large samples are better than small ones is that the bigger the sample, the smaller the bias.
   (b) If you have a sample of several measurements, all made under the same conditions, you can usually tell just from looking at the numbers whether the sampling process is biased, and whether the measurement process was biased.

(c) If you have a sample of several measurements, all made under the same conditions, you can usually get some idea just from looking at the numbers about how big the measurement errors are.

(d) An important reason why large samples are better than small ones is that chance-like errors tend to cancel each other when you compute averages.

(e) A statistic is a number you compute from a sample.

(f) The population average is an example of a statistic.

(g) Parameters are used to estimate statistics.

(h) For a random sample, each individual in the population has the same chance of being in the sample.

Answer

(a) F

(b) F

(c) T

(d) T

(e) T

(f) F

(g) F

(h) This is true for a *simple* random sample, or SRS.

3. For this exercise, the population is the box of numbered tickets shown below.

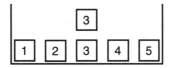

(a) The population average is equal to _____.

(b) If you draw a ticket at random and define chance error to be (# on ticket) − (population average), then there are five different possible values for chance error. List them.

(c) Which possible chance error is most likely?

(d) Suppose you get a sample by drawing three tickets at random, and that just by the luck of the draw, your sample turns out to be as follows.

| 1 | | 2 | | 3 |

    (i) The average for the sample is _____.

(ii) Is A or B correct?

A. The fact that the average for this sample is not equal to the population average proves that the sampling method is biased.

B. The difference between the sample average and population average is due to chance errors that come from the method of sampling.

Answer

(a) 3

(b) −2, −1, 0, 1, 2

(c) 0

(d) (i) 2

   (ii) B

4. Each of the following sampling methods is biased. For each, tell which individuals in the population are more likely to be chosen, which are less likely, and explain why.

(a) *Population:* all fish in Moosehead Lake, a large lake in northern Maine.

Sampling method: Drag a net with one-inch mesh (hole size) behind a motor boat, up and down the length of the lake.

(b) *Population:* all people over age 18 in New York City.

Sampling method: Run an advertisement in the *New York Times*, asking for volunteers; then choose at random from the list of people who volunteer.

(c) *Population:* all trees in a one-acre lot.

Sampling method: Assume you have a map of the lot showing the location of each tree. To get the first tree in your sample, you choose a point at random on the map and take the tree closest to that point for your sample. To get a second tree, you repeat the process by choosing a new point at random and taking the tree closest to that point. Continue choosing random points and closest trees until you have a sample of twenty trees.

(This method was once in common use until someone discovered the bias. If you don't see the bias at first, it might help to draw a little map with about ten trees. Put in several younger trees, which grow close together, and two or three older trees, whose large leaf canopies discourage other trees from growing nearby. Which trees are more likely to be chosen by the sampling method?)

Answer

(a) Fish small enough to get through the net won't be in the sample. Fish that tend to stay near the bottom in deep water will be unlikely to be caught.

(b) People who read the *New York Times* are not at all representative of all people over 18 in New York. Moreover, volunteers are rarely representative of the group that includes non-volunteers.

(c) Large trees, whose leaf canopy keeps other trees from growing nearby, will be overrepresented.

5. Give two reasons for randomizing the assignment of conditions to material.

Answer

Random assignment protects against bias, and allows you to rely on the predictable regularities of chance-like behavior when you analyze your results.

6. True or false: In a complete block design, the factor blocks is crossed with the factor of interest.

Answer True

7. A lot of the language of experimental design comes from agriculture. Here's a simple but typical example. You want to study the effect of two fertilizers, sulfate of ammonia (for nitrogen) and superphosphate (for phosphorus), on the yield of Brussels sprouts. You decide to use each fertilizer at four **levels**: 0, 30, 60, and 90 pounds per acre.

(a) List a complete set of treatment combinations in a two-way table.

(b) Suppose you have a bunch of similar square plots, 64 in all, with the same size and similar soil conditions. Which design principle would you use for assigning treatments to units?

Answer

(a)

Superphosphate (lbs/acre)

	0	30	60	90
Sulfate of ammonia (lbs/acre) 0				
30				
60				
90				

Each cell in the table corresponds to one treatment combination.

(b) You could use the principle of random assignment: randomly assign each treatment combination to four plots.

8. *Needle threading.* No one is going to win a Nobel prize for a study of the factors that affect the speed of needle threading. Nevertheless, hang on. This is an exercise that gives you a chance to apply many important principles of design, and you don't have to have the sort of specialized knowledge that you might need to plan other experiments.

It is reasonable to think that how fast you or I can thread a needle depends both on the color of the thread and on the background behind it. Black thread against a white background would certainly be easier than black thread against a black background. Suppose your goal is to measure the relationship between thread color (white, black, red) and background color (white, black) on the speed of needle threading.

(a) *Measurement.* There are many different ways to measure how fast a person can thread a needle; some of these ways are much better than others.

   (i) Think of two different ways, and describe each one as if you were writing out instructions for an assistant. Assume your assistant is a bit like a computer—very conscientious, but not terribly bright, so that you have to spell everything out in detail.

   (ii) Which of your two ways is better? Why?

(b) *Two-way completely randomized factorial design.* Tell how to run the experiment as a two-way factorial. List what you regard as any important advantages or disadvantages of the design.

(c) *Complete block design.* List any nuisance influences that you consider likely to be important. Pick the one you regard as most important, and describe two different ways to use that nuisance influence to define blocks as a nuisance factor.

(d) *Comparison.* Which of the designs in parts (b) or (c) do you judge better? Give a reason for your answer.

Answer

(a) Measurement.

  (i) *Method 1:* Timer says "Go" and starts a stopwatch. Subject threads needle and says "Done" when finished. Timer stops watch. *Method 2:* Get an electronic timer that rings a bell at the beginning and end of a 2-minute interval. When the bell rings, subject pulls a 20 inch length of thread through a needle as many times as possible before the bell rings again. Response = # times.

  (ii) Method 2 is better because it is less subject to errors from the reaction time of the person doing the timing.

(b) Two-way design. Randomly assign equal numbers of subjects to the six treatment combinations shown below:

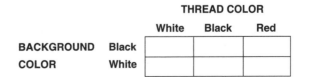

		THREAD COLOR		
		White	Black	Red
BACKGROUND	Black			
COLOR	White			

Advantages: (1) The design will give data that let you measure interaction. (2) Because you don't reuse subjects, your data will be free of effects due to practice or tiredness.

Disadvantages: (1) The design uses lots of subjects. (2) Residual errors will probably be large, because they include between-subjects differences.

(c) Complete block design. The most important nuisance influence is likely to be the subjects. Two ways to get blocks:

Sorting: Give subjects a pretest, and sort them into blocks of 6 according to how fast they can thread a needle.

Randomly assign treatment combinations to subjects within each block of 6.

Reusing: Get each subject to perform under all 6 conditions, randomizing the order of the conditions separately for each subject.

Advantages: Both CB strategies should lead to smaller residual errors than the CR (2) design because between-subjects differences will be estimated as block effects. Both designs let you measure interaction.

Disadvantages: The design based on sorting uses lots of subjects. The design based on reusing subjects requires many fewer subjects, but effects due to tiredness or practice will be included in the residual errors, making them bigger than they would be otherwise.

(d) Comparison. The block designs are both better than the completely randomized design because subject-to-subject variability is large. Of the two block designs, the one that reuses subjects is much less wasteful.

9. A study of oral contraceptives (birth control pills) from the 1970s gathered data on thousands of women. Among other things, the researchers recorded each woman's blood pressure and how many children she had. The study showed that if you compared blood pressures for women with two children and women with four children the average of those with four kids was about 30 points higher than for those with only two. The careless conclusion is that having kids causes your blood pressure to go up. Actually, though, selection bias is at work. Mothers in the first group (two kids) tend to be _____ than mothers in the second group, and as you get _____ , your blood pressure tends to go up. So for this data set, number of children is confounded with _____ .

Answer older, older, age

10. To study the effects of climate on health, you could compare the death rates of two states which have very different climates, like Florida and Alaska. (The death rate for a given year tells how many people out of every 100,000 died during that year.) For 1981, Florida's death rate was more than twice as high as Alaska's.

(a) The reason for using death rate instead of total number of deaths is to avoid confounding: if you looked at total number of deaths, the effect of climate would be confounded with _____: states that were _____ would tend to have higher values.

(b) Even death rate is not a good choice for a response, mainly because the effect of climate would still be confounded with _____. (Think about the kinds of people who move to the two states.)

Answer

(a) population; larger (more populous)

(b) age: the people in Alaska tend to be younger, the people in Florida tend to be older.

11. Suppose you wanted to do a small study of student attitudes on your campus. You could take a random sample, but it would be much easier to use a group of your friends as your sample.

(a) Which sample would give you a more uniform set of subjects, the random sample or the group of friends?

(b) Which sample would be more representative?

(c) For this study, if you have to choose between a uniform sample and a representative sample, which should you choose? Why?

Answer

(a) Groups of friends would probably be more uniform.

(b) The random sample would almost surely be more representative.

(c) The representative (random) sample is better because the results of your study would be more likely to apply to all students on campus.

12. If you have two sets of subjects available for a study, with the first set more uniform than the second, then (choose one of i–iii):

(a) Chance errors are likely to be smaller

    (i)   using the first set of subjects;

    (ii) using the second set of subjects;

    (iii) there is no way to tell.

(b) Bias is likely to be smaller

(i)   using the first set of subjects;
(ii) using the second set of subjects;
(iii) there is no way to tell.

Answer

(a) (i)
(b) (ii)

13. If you have two sets of subjects available for a study, with the first set more representative than the second, then
    (a) Chance errors are likely to be smaller
        (i)   using the first set of subjects;
        (ii) using the second set of subjects;
        (iii) there is no way to tell.
    (b) Bias is likely to be smaller
        (i)   using the first set of subjects;
        (ii) using the second set of subjects;
        (iii) there is no way to tell.

Answer

(a) (iii)
(b) (i)

14. *Hamlet,* Act II, Scene 2, Line 259: "There is nothing either good or bad, but thinking makes it so."

    Robert Rosenthal, who teaches psychology and social relations at Harvard, has done a large number of experiments that show how much our expectations influence what actually happens. One experiment looked at the influence of first-grade teachers' expectations on how well their students performed. All the first graders in the study were given a test, which their teachers were told was designed to predict which children would be likely to show a sudden spurt in how well they did in school. After the test had been scored, the teachers were given three lists of names, telling which ones of their students were likely to spurt ahead, which ones were likely to make average progress, and which ones probably would not show much intellectual growth.

What the teachers did not know was that the test was not a special predictor, but just a standard IQ test, and the three groups had nothing to do with the test: they were chosen using a chance device, and so were pretty much the same.

After a year the children were given another standard IQ test, and the differences in scores (second test minus first) were compared for the three groups. The results showed that how well the students did fit with what the teachers had been led to expect: IQ scores from the first group improved the most, and those for the third group improved the least.

(a) List the response and tell whether it is categorical or quantitative.

(b) List the conditions to be compared.

(c) Is the study experimental or observational?

Answer

(a) Response = change in IQ score; it is quantitative.

(b) The conditions are the teacher expectations.

(c) Experimental: conditions were assigned.

15. Three standard post-surgical treatments for breast cancer are chemotherapy, radiation therapy, and both. You can think of these as three out of four possible treatment combinations that come from crossing two sets of treatments. Show this by drawing and labeling a two-way table; put an X in the cell for the missing treatment combination. Why wouldn't you use a complete factorial design (all four combinations) here?

Answer

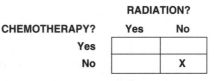

It would be unethical to give no treatment to patients who need it.

16. Leshner and Moyer investigated the relationship between the male hormone testosterone and two aspects of mouse behavior: aggressiveness, and how long it took the mice to learn to avoid attack. They compared three groups of male mice: castrated, sham operated (placebo), and castrated with testosterone supplements. (Their results showed that the hormone had a dramatic effect on aggressiveness, and essentially no effect on learning to avoid attack.)

This is another example of a two-way factorial plan with a missing cell. Show the factorial structure and missing cell with a table. Why were there no mice for the missing cell?

Answer

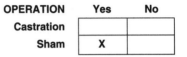

| | TESTOSTERONE REPLACEMENT | |
OPERATION	Yes	No
Castration		
Sham	X	

17. The two-way factorial design has served as the structure for many old-style studies in comparative physiology, which picked a bunch of animals (the conditions) and some physiological measure (the response) as a basis for comparing them. This idea can be extended to compare the way different species react to changes in their environment. Here's a simple example.

Suppose you want to use a two-way factorial design to compare the effects of temperature on warm-blooded and cold-blooded animals. Think about how you would do this, then answer the questions. (Be sure to include spiny anteaters among your subjects: these animals are in between warm and cold blooded.)

(a) Draw and label a two-way table showing your factors of interest and the combinations of their levels.

(b) Which of your factors of interest are observational, which are experimental?

(c) Consider the choice between the two-way plan and a different plan that reuses subjects. Under what circumstances would you prefer each one?

Answer

(a)

	AIR TEMPERATURE		
ANIMAL	40°F	60°F	80°F
Hamster			
Spiny anteater			
Bullfrog			

(b) Air temperature is experimental; animal species is observational.

(c) Use a completely randomized plan if you have many animals and little time. Otherwise, reuse animals, leaving adequate recovery time between treatments.

18. In order to assess the effectiveness of a new fertilizer, researchers applied the fertilizer to the tomato plants on the west side of a garden, but did not fertilize the plants on the east side of the garden. They later measured the weights of the tomatoes produced by each plant and found that the fertilized plants grew larger tomatoes than did the non-fertilized plants. They concluded that the fertilizer works.

(a) Was this an experiment or an observational study? Why?

(b) What are the explanatory and response variables in the study?

(c) This study is seriously flawed. Use the language of statistics to explain the flaw and how this effects the validity of the conclusion reached by the researchers.

(d) Could this study have used the concepts of "blinding" (i.e., does the word *blind* apply to this study)? If so, how? Could it have been double-blind? If so, how?

Answer

(a) This was an experiment, because the researchers applied the conditions (i.e., they created the two groups).

(b) The explanatory variable is amount of fertilizer (some or none). The response variable is tomato weight.

(c) The effect of east vs. west sunlight (and soil quality, irrigation, etc.) is confounded with the effect of the fertilizer. We don't know if the plants in the west part of the garden produced large tomatoes because of the fertilizer or because the west part of the garden gets more sunlight than the east part, has better soil, etc.

(d) Blinding could be used: the person weighing the tomatoes could be left unaware of which part of the garden they came from. Double-blinding does not apply here, however: it does not make sense to talk about the plants being blinded.

19. To compare surgical and non-surgical treatments of intestinal cancer, researchers examined the records of a large number of patients. On average, patients who received surgery lived much longer than those who did not receive surgery. The researchers concluded that surgery is more effective than non-surgical treatment.

(a) What are the explanatory and response variables in this study?

(b) On further investigation it is found that some patients had been assigned to non-surgical treatment because they were too ill to tolerate surgery. Use the language of statistics to explain how this affects the validity of the conclusion reached by the researchers.

(c) Suppose we wanted to conduct a new study of 60 patients in which we control for level of patient health (high, medium, low). Give a schematic outline of a randomized complete block design.

Answer

(a) The explanatory variable is type of treatment (surgical or non-surgical). The response variable is length of life following treatment.

(b) The effect of level of illness on longevity is confounded with the effect of surgery on longevity, so the conclusion is not valid. Surgery patients would probably have lived longer anyway.

(c)

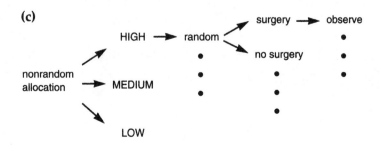

To conduct a randomized block design we first create blocks of patients and then randomly assign patients to treatment groups.

20. Erin Moore (Oberlin College '95) phoned a random sample of 25 male Oberlin students and asked them "Do you think that men should put the toilet seat down when finished?" She found that 68% said "yes."
  (a) Is 68% a parameter or a statistic? Why?
  (b) Explain, within the context of this setting, what is meant by response bias and how it might affect Erin's estimate.

Answer

  (a) The 68% value is a statistic, since it is calculated from the sample data. (The corresponding parameter would be the percentage of the entire population who would say yes.)
  (b) Men might lie, since they are being interviewed by a woman. This would introduce a positive response bias — a tendency for the statistic to overestimate the parameter. Thus, the percentage "yes" in the sample might be higher than the true percentage of men in the population who agree with the statement in the survey.

21. (Based on A.N. Tsoy et al., *European Respiratory Journal* 3 (1990): 235; via Berry, *Statistics: A Bayesian Perspective.*) Researchers wanted to compare two drugs, formoterol and salbutamol, in aerosol solution to a placebo for the treatment of patients who suffer from exercise-induced asthma. Patients were to take a drug or the placebo, do some exercise, and then have their "forced expiratory volume" measured. There were 30 subjects available.
  (a) Should this be an experiment or an observational study? Why?
  (b) Within the context of this setting, what is the placebo effect?
  (c) Briefly explain how to set up a randomized blocks design (RBD) here.
  (d) How would an RBD be helpful? That is, what is the main advantage of using an RBD in a setting like this?

Answer

(a) This should be an experiment, to eliminate confounding of the effects of age, severity of illness, etc. with the effects of the drugs.

(b) The placebo effect is improvement of forced expiratory volume that is caused by the psychological benefit of thinking that you are taking a helpful drug even if you are not.

(c) We could form 10 groups of 3 patients each according to age, sex, etc. Then within each group (block) we would randomly assign one person to placebo, one person to formoterol, and one person to salbutamol.

(d) With an RBD, we eliminate variability caused by age, sex, etc. This gives us more power to detect effects of the drugs.

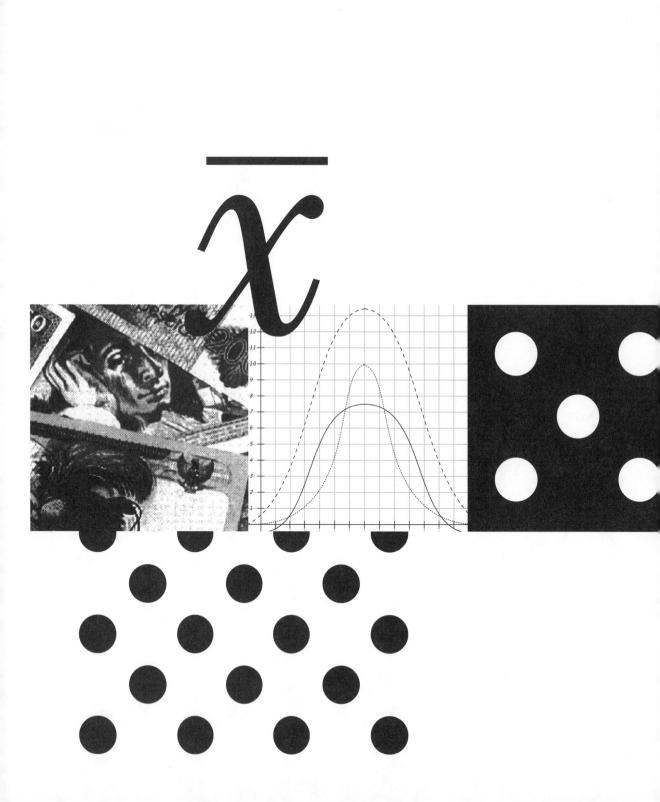

Probability Distributions

Summary _____

7.1 IN A NUTSHELL

Where does this topic fit in? Remember, from Topic 6, that data analysis consists of two phases: **exploration** and **inference.** The bridge that joins these two is built of probability-based data **production** (Topic 6) and the idea of imagining the results of a large number of **repetitions.** Probability distributions give us a way of summarizing the results of the large number of repetitions. Probability is a huge and fascinating branch of mathematics, but also very hard to learn to think about intuitively. For a first statistics course, however, the probability you need is limited to answering a specialized kind of question: "Suppose I produce data using one of the probability-based methods of Topic 6, and compute a summary statistic like the sample mean. If I were to repeat that process a large number of times—generate the data and compute the summary—how would the summary behave?" The answer to this crucial question is given by a sampling distribution, which is a particular kind of probability distribution.

What is a probability distribution? You've actually seen ideas very similar to this in Topic 3, on the normal distribution for data. The 68/95/99.7 rule is a summary for a family of **data distributions;** the rule tells us about the values we can get, and how often they occur. The only difference between a data distribution and a **probability distribution** is that the data distribution gives *observed* percentages, and tells how often the data values *actually* occurred; a probability distribution gives *theoretical* percentages, or chances, which tell how often the values *can be expected* to occur in a large number of repetitions.

Four ways to think about probability distributions Because these ideas are both hard and important, it will repay your effort to work at understanding them. It helps to try to become comfortable with four different variations on the one basic idea. You can think of a probability distribution as follows:

- *numerically,* as a table listing all possible outcomes, together with the chance for each outcome;
- *visually,* as a probability histogram, with one bar for each outcome, and the area of the bar proportional to the chance of the outcome;

- *physically*, as a "box model," with numbered tickets as the outcomes and the chance of each outcome given by the fraction of tickets in the box having that number on them; and
- *predictively*, as an answer to the question, "What will happen if I repeat a particular chance process a very large number of times?"

It may help to notice that a distribution is nothing more than a way of organizing and thinking about a list of numerical outcomes. Suppose, for example, your "box model" contains six tickets: 0, 1, 2, 1, 2, 2. You can summarize the list by telling how many times each number occurs in the list: 0 (1 out of 6), 1 (2 out of 6) and 2 (3 out of 6). In essence, that's all a distribution is; the rest is a matter of learning how to think about and use distributions.

Summarizing probability distributions As you know from your work with data distributions, the great value of a plot is that it helps you summarize a data set by describing its shape in words and giving a couple of numerical summaries. The same is true of probability histograms. Often, we can summarize them by telling the shape, locating the center, and measuring the spread. If the shape is roughly normal, then we locate the center by telling the expected value, and measure the spread using the standard deviation.

7.2 DISCRETE PROBABILITY DISTRIBUTIONS

Expected value There are four useful ways to think about the **expected value** of a discrete distribution, one for each of the four ways to think about the distribution itself:

- *numerically*, as the sum of outcomes times chances;
- *visually*, as the point (numerical value) where the probability histogram balances;
- *physically*, in terms of a "box model," as the average of the tickets in the box;
- *predictively*, as an answer to the question, "If I repeat the chance process a very large number of times, what will be the long-run average of the outcomes?"

Standard deviation The **standard deviation** for a discrete probability distribution is very similar to the standard deviation for data; both measure the typical size ("standard") of a deviation using the root mean square as a kind of average. As before, you read "SD = root mean square deviation" backwards:

- *deviation* = the distance from an outcome to its expected value: $y - EV(y)$;
- *square* the deviations (which makes the negative deviations positive);
- *mean:* take the mean of the squared deviations—multiply by the chances, and add, and finally;
- *root:* take the square root, to undo the effect of the squaring.

Binomial distributions The **binomial distributions** are among the most useful of the discrete probability distributions. You can use them to describe any real-world process with the following four properties.

B inary outcomes: There are only two kinds of outcomes, like heads and tails, died and survived, defective and non-defective, success and failure.

I ndependent outcomes: Each outcome is determined in a way that doesn't depend on previous outcomes, and doesn't influence future outcomes.

N umber of outcomes, n is fixed in advance. (If you toss a coin 10 times, n is fixed in advance; if you toss until you get 5 heads, n is not fixed, since it depends on the outcomes.)

S ame probability (p) of success for all the outcomes.

The binomial outcome is the number of successes in the set of n individual "trials." (Such individual trials, e.g., a single coin toss) are sometimes called **Bernoulli trials.**) For a binomial distribution, the chance of exactly r successes in n trials is

$$P(r) = \binom{n}{r} p^r (1-p)^{n-r}$$

The expected value is np = (number of trials) × (chance of success on each trial), and the standard deviation is $\sqrt{np(1 - p)}$.

Continuous probability distributions Unless you have studied calculus, you will probably find it easiest to think about a continuous distribution as a smooth approximate version of a discrete probability histogram. This allows you to rely on what you know about discrete probability distributions. The expected value and standard deviation for a continuous distribution behave pretty much as they do for discrete distributions. In particular, the expected value is the balance point for the probability histogram, and the standard deviation is the root mean square deviation from the *EV*. By far the most important continuous distributions are the **normal distributions,** which are introduced in Topic 3 in the closely related form of distributions for data. Normal distributions are important for inference because, as you'll see in the next topic, many of the summary statistics you compute from data will tend to follow a normal distribution, at least approximately, provided you use the probability-based data production methods of Topic 6.

Self-Testing Questions

1. You are responsible for serving New Year's dinner when New Year's Day falls on a Friday. Assuming each day has an equal chance of falling on New Year's, what are the expected number of times you will be serving New Year's dinner the twenty years out of the next thirty your family will be able to get together? (Don't laugh. Some families actually work like this.) Round your answer to the nearest whole number.

➤ *Solution*

Approximately 3, since $20 \times \frac{1}{7} = 2.86$

2. What is the standard deviation of the random variable in Question 1? Round your answer to the nearest hundredth.

Answer

$$\sqrt{20\left(\frac{1}{7}\right)\left(\frac{6}{7}\right)} = 1.56$$

3. A stoplight has a cycle of 20 seconds for green, five seconds for yellow, and 30 seconds for red. This means that you have a $\frac{35}{55} = \frac{7}{11}$ chance of being stopped for a traffic light, assuming you stop for yellow lights. On a trip, you know you will go through 27 traffic lights. What probability distribution gives the probability that you will be stopped by n traffic lights along the way?

Answer

$$\text{Bin}(27, {}^{7}\!/\!{}_{11}) = \binom{27}{n}\left(\frac{7}{11}\right)^{n}\left(\frac{4}{11}\right)^{27-n}$$

4. What is the expected number (to the nearest whole number) of traffic lights that you will get stopped at?

Answer

$$({}^{7}\!/\!{}_{11}) \times 27 = 17$$

5. What is the standard deviation of this distribution?

Answer

$$\sqrt{(27) \times \left(\frac{7}{11}\right) \times \left(\frac{4}{11}\right)} = 2.5$$

6. A mass mailing company claims to have a response rate of 37%. In a mailing of 1000 pieces, the response size was 310. How many SDs is this from the expected return?

➤ **Solution**

The expected return is 370. The SD of this distribution is $\sqrt{1000 \times .37 \times .63}$ or approximately 15. Thus, the response is $370 - 310 = 60 = 2$ SDs from the expected value.

7. True or false: You are playing a game in which you pull cards one at a time from a standard pack of 52 without replacement. After pulling ten times, the probability that you pulled 5 heart cards can be found using the binomial distribution.

➤ **Solution**

False, the probability could be found using the binomial only if cards were pulled and then put back in the deck.

8. The following ticket numbers are in a box: 1, 1, 2, 2, 3, 4, 4, 5, 5. One ticket is chosen at random.
(a) What is the expected value of this outcome?
(b) What is the probability of getting a number greater than 3?

Answer
(a) 3
(b) 4/9

9. In 1987, .018 of all vehicles involved in motor accidents were motorcycles. Suppose we take a random sample of 1000 vehicles that were in accidents and we count the number of motorcycles. Can we use a binomial model for this situation?

Answer Yes

10. Suppose we sample people one at a time until we get exactly 5 who have type AB blood. Can we use a binomial to model how many people we will need to sample to get 5 with type AB blood?

➤ *Solution*

No; n is not fixed in advance.

11. Suppose a baseball player has a .03 chance of hitting a home run (4 bases), a .01 chance of hitting a triple (3 bases) and a .03 chance of hitting a double (2 bases), a .250 chance of hitting a single or a walk (1 base), and otherwise makes an out. (These probabilities are based on data from the 1996 National League Championship Series.)
(a) What is the chance the player will make an out?
(b) What is the chance the player will get at least 2 bases?

Answer
(a) .68
(b) .07

12. Make a box model using the probabilities given in Question 11. What is the expected value of the number of bases a player will get? (This is called the slugging percentage.)

➤ *Solution*

A box with 100 tickets should have 3 tickets labeled "4," 1 labeled "3," 3 labeled "2," 25 labeled "1," and 68 labeled "0." The expected value is

$$\frac{3 \times 4 + 1 \times 3 + 3 \times 2 + 25 \times 1}{100} = \frac{46}{100} = .46$$

13. Consider the probabilities from Question 11. What is the SD of the distribution of bases?

➤ *Solution*

Since the *EV* is .46, the deviations are –3.54, –2.54, –1.54, –.54, and .46. Squaring the deviations and multiplying each by the corresponding probability (from Question 11) gives us .7284. The SD is $\sqrt{.7284} = .853$.

14. In the United States, 44% of the population has type O blood. Suppose we take a sample of 25 people. How many of them would we expect to have type O blood?

Answer 11

15. What is the standard deviation of the number of people with blood type O in a sample of size 25, if the probability of type O blood is 44%?

Answer

$$\sqrt{25 \times .44 \times .56} = 2.48$$

16. In the United States, 10% of the population has type AB blood. If we take a sample of 2 people,
(a) What is the chance both will have type AB blood?
(b) What is the chance neither will have type AB blood?
(c) What is the chance exactly 1 of them will have type AB blood?

Answer
(a) .01
(b) $.9^2 = .81$
(c) $1 - .01 - .81 = .18$

17. The number of mistakes per page that a typist makes is given in the following table.

Mistakes	Frequency
0	.37
1	.37
2	.18
3	.06
4	.02

What is the probability that a randomly chosen page will have more than 1 mistake?

Answer .26

18. Consider the typist from Question 17. What is the expected number of mistakes per page for this typist?

Answer .99

19. Consider the typist from Question 17. What is the standard deviation of the number of mistakes per page for this typist?

Answer .985

20. Of the Ph.D.s awarded in mathematics to U.S. citizens, 28% are earned by women. If we take a random sample of 5 mathematics Ph.D.s what is the chance that all 5 will be men?

Answer $.72^5 = .19$

21. The English Department at Oberlin College has 20 faculty members. Of these, 13 are men and 7 are women. The Dean wants to create a committee of 3 English faculty by choosing 3 names at random. However, the Dean is worried about gender balance on the committee. Make a box model to represent the English Department. How many tickets go into the box? What kinds of tickets are there? How many are there of each kind?

Answer

The box should have 20 tickets. There are two kinds of tickets, 13 that say "man" and 7 that say "woman."

22. Consider the situation in Question 21. What is the probability that all 3 faculty chosen for the committee will be men?

➤ *Solution*

We need to draw from the box without replacement. To find the chance of getting 3 men we need to find the chance that the first person chosen is a man; this is $^{13}/_{20}$. Then we need to find the chance that the second selection to be a man given that the first is a man; this is $^{12}/_{19}$, since there are 19 faculty left and 12 of the 19 are men. Finally, we need to find the chance that the third selection is a man, given that the first two were men; this is $^{11}/_{18}$. The probability of getting 3 men is therefore $^{13}/_{20} \times ^{12}/_{19} \times ^{11}/_{18} = ^{1716}/_{6840} = .251$.

23. Only 40% of the voting-age population votes in a typical Presidential election. Suppose we take a random sample of 2 people of voting age and find out whether or not they voted in the last election.
   (a) What is the probability that both of them will have voted?
   (b) What is the probability that at least 1 of them voted?

➤ *Solution*

(a) $.5^2 = ^1/_4 = .25$

(b) .75. The probability that exactly 1 voted is $2 \times .5 \times .5 = .5$. That is, the probability that exactly 1 voted is P(1st voted, 2nd did not) + P(1st did not vote, 2nd did vote) $= .5 \times .5 + .5 \times .5 = .5$. The probability that both voted is .25. Thus, P(at least 1 voted) = P(exactly 1 voted) + P(both voted) $= .5 + .25 = .75$.

24. Consider the setting of Question 23. Suppose we know that at least 1 of the 2 people voted. What is the probability that both of them voted given that at least 1 of them voted?

➤ *Solution*

This is a conditional probability. *P*(both voted given at least one voted) is *P*(both voted and at least one voted)/*P*(at least one voted) = *P*(both voted)/*P*(at least one voted) = .25/.75 = 1/3. Note that the event "both voted and at least one voted" is the same as the event "both voted."

25. To play the "4 spot" version of the 80¢ Keno game at the Golden Strike Inn Casino in Las Vegas you pay 80¢, choose 4 numbers from 1 to 80, and hope that your numbers are winners. You have probability .003 of winning $120.00, probability .043 of winning $2.00, probability .213 of winning $0.80, and probability .741 of winning nothing.
(a) What is the expected value of your winnings?
(b) Subtract $0.80 from your answer to part (a) to find your expected net winnings.

Answer

(a) The expected value is .003 × 120.00 + .043 × 2.00 + .213 × 0.80 + .741 × 0 = 0.36 + 0.086 + 0.1704 + 0 = 0.6164.
(b) The expected net winnings are 0.80 − 0.6164 = −0.1836, or roughly −18 cents.

26. (Based on an article in the 10/31/95 edition of *The Plain Dealer* with the headline "Scientists find sign of homosexuality gene.") Scientists examined a particular region of the X chromosome in 32 pairs of homosexual brothers. In 22 of the 32 pairs, the brothers "shared the same version of the genetic material"; let us call this "producing a match." Suppose that there is no genetic link at work so that, in fact, the chance of brothers producing a match is ½. Find the probability that at least 22 of 32 pairs would produce a match. Use the normal approximation with continuity correction.

Answer

The number of matches has a binomial distribution with mean 32 × (½) = 16 and standard deviation

$$\sqrt{32 \times \frac{1}{2} \times \frac{1}{2}} = 2.83. \text{ The probability of at least 22}$$

matches is $P(Z > \frac{21.5 - 16}{2.83}) = P(Z > 1.94) = .0262$.

Note that we use 21.5 in place of 22 as a continuity correction.

27. For each of the following situations, state whether or not a binomial would be an appropriate probability model for the variable Y. If the binomial model is not appropriate, state why not.

(a) Ten percent of the people in the United States have type B blood. Suppose you take a random sample of 5 families that have 4 members each and you determine the blood type of each person. Let Y be the number, out of 20, who have type B blood. Is Y a binomial random variable? If not, why not?

(b) Seeds of the garden pea (Pisum sativum) are either yellow or green. A certain cross between pea plants produces progeny that are in the ratio 3 yellow:1 green. Suppose your goal is to get 3 yellow, but you don't care how many green you get. So you sample, one at a time, until you have exactly 3 progeny that are yellow. Let Y be the number of progeny you have to observe in order to get 3 yellow. Is Y a binomial random variable? If not, why not?

(c) Some students eat at a college dining hall, some eat in a CO-OP, and some eat elsewhere. Suppose you take a random sample of 20 students and you ask each of them where they eat dinner. Let Y be the number of students out of 20, who eat at a college dining hall. Is Y a binomial random variable? If not, why not?

Answer

(a) Y is not a binomial random variable because the blood types of members of a family are not independent of one another.

(b) Y is not a binomial random variable because the sample size is not fixed.

(c) Y is a binomial: the outcomes are binary (the student either eats at the dining hall or does not), the trials are independent (it is a random sample), the sample size is fixed at 20, and the probability of a success is the same for each trial (each randomly chosen person has the same chance of eating in the dining hall).

X

3	276.7	WaPost h	4.80	17	92	335.50 −2.13
1	5	WaWater	1.24	14	326	18.50 −.13
12.13	7.50	WastMln	..	..	1417	7.63 −.13
33.63	18.38	Waters	..	51	501	30.38 +.50
43.75	17.00	Watkln	.48	73	433	26.25 −.25
34.38	11.31	Watsco s	.14	33	116	29.50 −.63
24.63	15.50	Watts h	.31	..	363	24.38 +.25
6.63	1.00	Waxmn	..	3	786	5.88 −.13
38.13	23.13	WthfrdE	..	..	1012	35.63 −.13
20.00	15.25	WebbD	.20	..	274	16.50 ..
36.50	23.50	Weeks n h	1.72	34	763	36.00 ..
44.75	34.25	WeinRl	2.48	22	374	42.88 −.25
4.38	2.00	Weirt	..	..	901	3.00 ..
34.88	28.75	WeisMk	.92	16	68	30.75 −.38
24.88	15.88	Wellmn	.32	20	666	17.63 −.25

Sampling Distributions

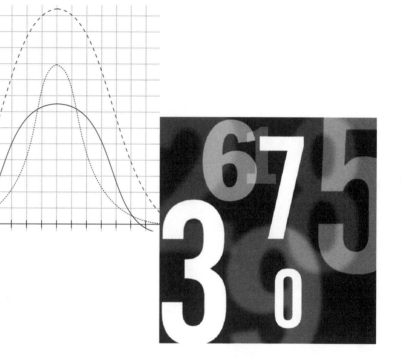

Summary _____

8.1 IN A NUTSHELL

Where does this topic fit in? Remember from Topic 6 that data analysis consists of two phases: **exploration** and **inference**. The bridge that joins these two is built from probability-based data **production** methods (Topic 6) and the idea of imagining the results of a large number of **repetitions**. Probability distributions (Topic 7) give us a way of summarizing the results of the large number of repetitions. **Sampling distributions** are probability distributions for numerical summaries (statistics) computed from simple random samples or from other probability-based methods of data production. They answer the question, "How would the summary statistic behave if I were to repeat the whole process (produce the data and compute the summary) a large number of times?" In practice, the two most useful sampling distributions are for the sample proportion and the sample mean. We use sampling distributions to justify such statements as "For 95% of all random samples, the observed average will be within .01 of the true population mean." The example that follows illustrates how you can use a sampling distribution for inference.

Example Many years ago, "San Diego Reader" wrote,

Dear Abby: You wrote in your column that a woman is pregnant for 266 days. Who said so? I carried my baby for ten months and five days, and there is no doubt about it because I know the exact date my baby was conceived. My husband is in the Navy and it couldn't have possibly be conceived at any other time because I only saw him once for an hour, and I didn't see him again until the day before our baby was born.

I don't drink or run around, and there is no way this baby isn't his, so please print a retraction about that 266-day carrying time because otherwise I am in a lot of trouble.

The book[1] where I found this example asks the question, "Is San Diego Reader lying?" and uses it to illustrate the logic of inference as follows. Given an observed summary statistic, $y = 310$ days (ten months and five days), the duration for a randomly chosen pregnancy, has a probability distribution that is approximately normal, with $EV = 266$ and $SD = 16$. So we can ask,

---

[1] Richard J. Larson and Donna Fox Stroup (1976). *Statistics in the Real World*, pp. 52–57. New York: Macmillan.

"Suppose I were to repeat the process (chose a pregnancy at random and record its duration) a large number of times. What's the chance I'd get a duration of 310 days or longer?" If we express 310 in standard units (Topic 3), we get 2.33, and we can use a normal probability table (also Topic 3) to find the chance, which turns out to be just under .01, or less than one time out of 100. According to the logic of inference, either a very surprising event has occurred, or else San Diego Reader was not telling the truth, and in fact her pregnancy lasted less than 310 days. A person who carelessly applied the methods of inference would conclude that she was lying.

However Although the example does illustrate how sampling distributions are used for inference, it also illustrates how *people can jump to the wrong conclusions by ignoring data production*. The logic of inference starts from the assumption that your data come from probability-based production methods. But San Diego Reader was *not* chosen at random from the population. The sample was a voluntary response sample (Topic 6), and it is reasonable to suppose that she wrote to Abby precisely because of the unusual length of her pregnancy. According to the sampling distribution, roughly one of every 100 pregnancies lasts as long as 310 days, so even though the chance is tiny, in absolute terms there must be a very large number of pregnancies this long. So while it would be very surprising if one pregnancy, chosen at random, lasted 310 days or more, it is not at all surprising that at least one of *Dear Abby's* thousands of readers was pregnant for 310 days. The moral: *If you want to use methods of inference, you must use probability-based methods of data production*.

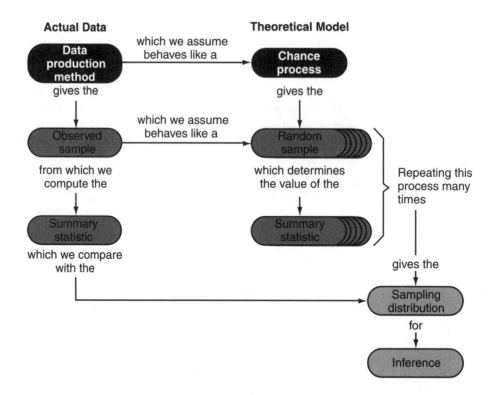

8.2 SAMPLING DISTRIBUTIONS

Finding a sampling distribution To find a sampling distribution, start with a known population and create all possible samples. For each sample, compute the value of, say, the sample mean. The set of all possible values of the mean, together with the corresponding chances, is its sampling distribution. You can think of this in four steps.

1. Start with a known population.
2. List all possible random samples.
3. For each sample, compute the value of your summary statistic.
4. Count the values to find the chances for each value. (For example, if there are 25 possible samples, and 10 of them give a sample mean of 3, then the chance of 3 is $10/25 = .4$.)

Summarizing a sampling distribution To use a summary statistic like the sample mean or sample proportion for inference, we need three pieces of information about its sampling distribution: the center (*EV*, or expected value), spread (*SE*, or standard error), and shape. The **expected value** of a sampling distribution tells us the population value that the statistic tends to be near. The **standard error** of a statistic (defined as the standard deviation of its sampling distribution) tells you how far (root mean square distance) the statistic tends to be from its expected value. If the shape of the sampling distribution is roughly normal, then we can use the 68/95/99.7 rule (Topic 3) to get approximate chances.

Expected value and bias The expected value of a statistic is both the balance point of the probability histogram of its sampling distribution and the long-run average value in repeated samples. If this expected value is equal to the population parameter you want to estimate, then the statistic is **unbiased**. The **bias** of a statistic is the distance from its expected value to the population parameter. Sometimes bias is due to a poor choice of summary statistic. More often, bias is due to a poor method for choosing the sample.

The sample proportion If you have a binomial process (Topic 7), and you use your sample to compute the proportion of "successes" $\hat{p}$, then your sampling distribution will have

- *center:* $EV = p$ (the sample proportion is an unbiased estimator of the population proportion),
- *spread:* $SE = \sqrt{p(1-p)/n}$ (the spread is proportional to "1 over root n," which means that the larger your sample, the less variable your summary statistic will be), and
- *shape:* the larger your sample, the more nearly normal will be the shape of the sampling distribution.

Qualifications: First, if you are sampling without replacement—the usual situation—the formula for the SE is an approximation, and you should use it only if your sample is no bigger than 10% of the population. Second, the approximate normal shape depends both on the sample size n being large and on the population proportion p being far enough from the extremes of 0 and 1. If both np and $n(1-p)$ are 10 or more, the normal approximation will work well.

The sample mean If you have a quantitative response, and you use your sample to compute the mean response value $\bar{y}$, then your sampling distribution will have

- *center:* $EV = \mu$ (the sample mean is an unbiased estimator of the population mean),
- *spread:* $SE = \dfrac{\sigma}{\sqrt{n}}$ (the spread is proportional to "1 over root n," which means that the larger your sample, the less variable your summary statistic will be), and
- *shape:* the larger your sample, the more nearly normal will be the shape of the sampling distribution.

Qualifications: First, if you are sampling without replacement—the usual situation—the formula for the *SE* is an approximation, and you should use it only if your sample is no bigger than 10% of the population. Second, if your population is normal, the sampling distribution will be normal also. If your population is not normal, then the sampling distribution depends both on the sample size n being large and on the population shape being far enough from extremes shapes, such as extreme skewness or very long tails. If the population shape is not normal but is not extremely skewed, then a sample size of 30 is usually large enough to make the normal approximation work well.

Using the normal approximation Suppose the normal approximation applies to your sampling distribution. Then based on the 68/95/99.7 rule, you can say that, unless you were unlucky and got a sample that only occurs 5% of the time, the value of your summary statistic is within 2 *SEs* of the true parameter value. Moreover, unless you were extremely unlucky, and got a sample that occurs only 3 times in every 1,000 samples, the value of your summary statistic is within 3 *SEs* of the unknown true value.

Why is this topic (sampling distributions) so hard? Many students and teachers of statistics regard this as the hardest topic in a first course. There are two reasons. First, the sampling distributions that get used for real applications require you to work with computing rules that you're still in the process of learning. This means that a large part of your attention still has to go to computing rules, making it harder to pay attention to the ideas. Second, the ideas themselves are hard. They involve three similar-seeming but in fact fundamentally different sets of numbers, each set with a very different role and meaning—the population, the sample, and the set of values of the statistic.

What can you do about it? A good strategy is for you to do a physical simulation of the sampling process, using a set-up so simple that you don't have to pay much attention to the arithmetic. A "hands-on" experience of the sort that follows may strike you as too simple-minded to be useful, but give it the few minutes it will take you just to try it out. It's a lot easier to master the ideas—how the three sets of numbers (population, sample, and values of the statistic) are related and how to keep them straight in your mind—if you can refer to a concrete example. Here's the simulation.

Step 1. *Population:* Get three or four pennies with different dates, and use the last digit of the date as your variable. For example, I just took four pennies from my pocket, and their dates are 1994, 1995, 1996 and 1996, so my population is 4, 5, 6, 6.

Step 2. *Sample:* Shake up the pennies in your hand and drop two on your desk—that's your sample. (I just did that, and got 5, 6 as my sample.)

Step 3. *Summary statistic:* Take the average of the values in the sample. (For the sample 5, 6, the average is 5.5.)

Step 4. *Sampling distribution:* Put your sample pennies back in the population and repeat Steps 2 and 3. Keep track of all your sample averages using a dot graph. That graph represents the sampling distribution (of the mean of samples of size 2 for your particular population.)

Variations Notice that you get a new sampling distribution if you change any one of the population, sample size, or summary statistic. You may not want to try variations, but it is worth taking a few minutes to think about them.

1. For example, if you have a population of size 4, as in my example above, and you use the sample mean, imagine four sampling distributions, one for samples of size $n = 1$, a second for $n = 2$, a third for $n = 3$, and finally one for $n = 4$. How will these four distributions compare?

2. Now go back to samples of size $n = 2$, and consider the sampling distributions for different summary statistics: the sample mean, sample median, minimum, maximum, and sample range. What can you say about how the corresponding five sampling distributions will compare?

Self-Testing Questions

1. Fill in the blank:
   (a) If you increase the number of observations that go into an average, the standard error for the average goes _____ (up/down).
   (b) If you compare two averages, one based on 10 observations with a small standard deviation, the other based on 10 observations with a large standard deviation, the _____ (first/second) will have the smaller standard error.

   Answer
   (a) down
   (b) first

2. The "root n property." (Fill in the blanks.) Assume SD = 1.
   (a) If the number of observations in a group goes from 25 to 100, then the *SE* for the average of the observations goes from _____ to _____.
   (b) If the number of observations in a group goes from 9 to 25, then the *SE* for the average of the observations goes from _____ to _____.
   (c) If the number of observations in a group goes from n to $4n$, then the *SE* for the average of the observations goes from _____ to _____.
   (d) If the number of observations in a group gets multiplied by n, then the *SE* for the average of the observations gets multiplied by _____.

   Answer
   (a) 0.2, 0.1
   (b) $\frac{1}{3}, \frac{1}{5}$
   (c) $\frac{1}{\sqrt{n}}, \frac{1}{2\sqrt{n}}$
   (d) $\frac{1}{\sqrt{n}}$

3. A normal probability histogram is shown below. Use it to match the following intervals of outcomes with their chances.

An Outcome
(a) Greater than 0
(b) Between −1 and 1
(c) Bigger than 3

Chance
(i) .135%
(ii) 67%
(iii) 50%

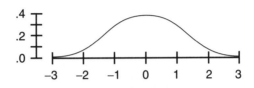

➤ *Solution*

(a) Greater than 0. The histogram is symmetric with exactly half its area on either side of 0. The chance is 50%

(b) Between −1 and 1. The area above this interval is roughly two-thirds of the total: chance = 67%.

(c) There is hardly any area to the right of 3: chance = .135%

4. Match each mechanism and summary in (a)–(d) with its histogram (i)–(iv).
(a) Mechanism: Toss 2 coins Summary: # Heads
(b) Mechanism: Toss 4 coins Summary: (# Heads)/2
(c) Mechanism: Draw twice Summary: Average of the
                from   draws

(d) Mechanism: Same as   Summary: SD for the draws
                for part(c)

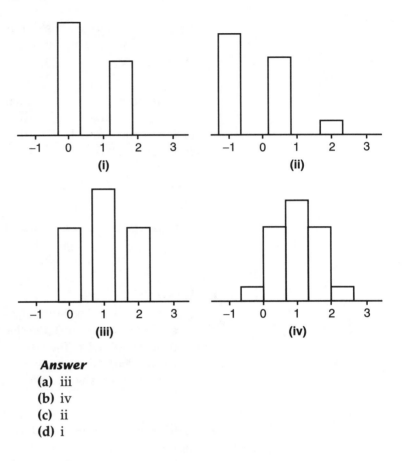

Answer
(a) iii
(b) iv
(c) ii
(d) i

5. The purpose of this exercise is to illustrate two facts: That if your population is small, there is a big difference between sampling with and without replacement, but that if your population is large, the two sampling methods behave pretty much alike.

(a) Suppose you draw two tickets at random from a box with just two tickets, labeled 1 and 2. As you think about the list of all possible samples, consider samples like (First draw = 1, Second draw = 2) and (First draw = 2, Second draw = 1) to be different.

(i) How many different samples (of size 2) are possible if you sample without replacing tickets? List the samples.

(ii) If you sample with replacement? List the samples.

(iii) If all possible samples in part (ii) are equally likely, what is the chance that your sample contains the same individual twice?

(b) Suppose your population has three different tickets. (1, 2, 3) and you draw a random sample of size 2. Answer (ii)–(iii) on p. 212.

(c) Suppose your population has four different tickets, and you draw a sample of size 2. Answer part (iii) on p. 212.

(d) Suppose your population has N tickets. Answer (iii) on p. 212. Use your answer to part (iii) to find the chance of getting the same individual twice for populations of size 5, 10, 50, 100, 1000, and 1,000,000.

Answer

(a) (i) Two samples: (1, 2) and (2, 1).

   (ii) Four samples: (1, 1), (1, 2), (2, 1), and (2, 2).

   (iii) 0.5 (2 out of 4)

(b) There are 9 possible samples. Three of these contain the same individual twice, so the chance is $3/9 = 1/3$.

(c) The chance is $4/16 = 1/4$.

(d) For a population of size N, the chance of getting the same individual twice is $1/N$.

6. In October of 1996, a random sample of 600 likely voters in Ohio found that 45% intend to vote for Clinton. A SRS of 600 likely voters in Lorain County found that 44% intend to vote for Clinton.

(a) Which of these results, if either, has a smaller margin of error?

   (i) The Ohio result;

   (ii) The Lorain County result;

   (iii) They have the same margin of error. Why?

(b) Is the 45% value for Ohio a parameter or a statistic? Why?

Answer

(a) (iii) The margin of error is the same in each case. The margin of error depends on the sample size, which is 600 in each case, but not on the population size (provided the population is much larger than the sample, which is true here).

(b) The 45% value is a statistic, since it is calculated from the sample data. (The corresponding parameter would be the percentage of the entire population who intended to vote for Clinton.)

7. In the United States, 45% of the population is age 35 or over; call these persons "old." You are to take a random sample and may choose a sample size of either 20 or 100. You will win $100 if the sample percentage of "old" persons is between .4 and .5. Should you choose a sample size of 20 or of 100? Why?

Answer

You should choose a sample size of 100. As the sample size goes up, the standard error of the sample percentage goes down. Thus, you are more likely to get a sample percentage close to 45% (which is the population percentage) if the sample size is large than if the sample size is small.

8. Kathy Konstantatos (Oberlin College '95) took a random sample of coffee drinkers and found that 6 of 50 (12%) drink decaffeinated coffee. In the context of this setting, explain what is meant by the sampling distribution of a percentage.

Answer

If we take repeated samples of size 50 coffee drinkers, the percentages of decaffeinated coffee drinkers that we obtain vary from one sample to the next. The sampling distribution of the percentage is the distribution that these sample percentages follow.

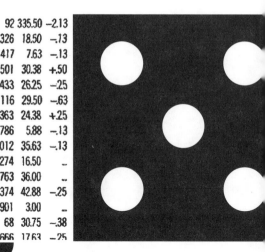

3	276.7	WaPost h	4.80	17	92	335.50	−2.13
15	75	WaWater	1.24	14	326	18.50	−.13
12.13	7.50	WastMln	..	..	1417	7.63	−.13
33.63	18.38	Waters	..	51	501	30.38	+.50
43.75	17.00	Watkln	.48	73	433	26.25	−.25
34.38	11.31	Watsco s	.14	33	116	29.50	−.63
24.63	15.50	Watts h	.31	..	363	24.38	+.25
6.63	1.00	Waxmn	..	3	786	5.88	−.13
38.13	23.13	WthfrdE	..	..	1012	35.63	−.13
20.00	15.25	WebbD	.20	..	274	16.50	..
36.50	23.50	Weeks n h	1.72	34	763	36.00	..
44.75	34.25	WeinRl	2.48	22	374	42.88	−.25
4.38	2.00	Weirt	..	..	901	3.00	..
34.88	28.75	WeisMk	.92	16	68	30.75	−.38
24.88	15.88	Wellmn	.32	20	666	17.63	−.25

topic 9
Introduction to Inference

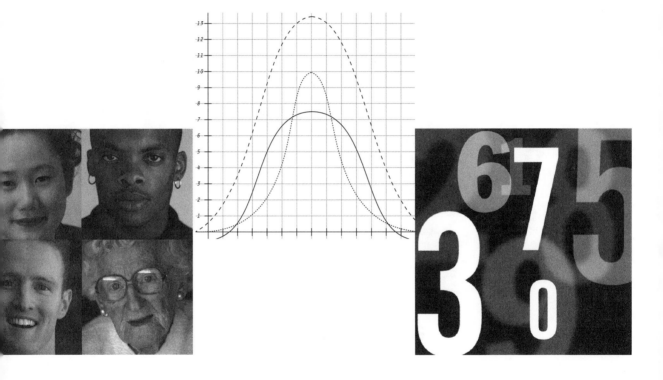

Summary

9.1 IN A NUTSHELL

Where does this topic fit in? Remember from Topic 6 that data analysis consists of two phases: **exploration** (finding and describing patterns) and **inference** (drawing conclusions). Exploration deals with the data you actually have—what you see is all there is. Inference, on the other hand, uses the data you see to try to answer questions about the data you might have gotten, or in other words, to answer questions about the probability-based process that produced your data. If you have a random sample from a population, inference uses the sample to draw conclusions about the population. If your data come from a randomized experiment, inference uses the observed results to try to generalize about what the effects of your treatments would be if you were to repeat the experiment a large number of times.

At this point, you should try to concentrate on learning the basic ideas. Specific methods for inference come in later topics.

Confidence intervals Informally, **confidence intervals** answer the question, "What values of the unknown parameter are reasonable, given the observed data?" To construct a confidence interval, you start with an estimate computed from the sample. The interval takes the form

$$\text{estimate} \pm \text{margin of error,}$$
$$\text{with margin of error} = (\text{critical value}) \times (SE \text{ of estimate}).$$

The **critical value** comes from a computer or from a table of values for the relevant sampling distribution, and is chosen to give a particular confidence level. The **confidence level** is the fraction of random samples for which the interval contains the true value of the parameter.

Significance testing Informally, **significance tests** answer the question, "Is such-and-such a reasonable value for my unknown parameter, given the observed data?" The simplest tests have a three-part structure:

- the **null hypothesis** H_0, which specifies a value for the population mean, or whatever parameter you are interested in,
- a **test statistic,** computed from the data, and which typically tells how many SEs it is from an estimate based on the data to the parameter value specified by the null hypothesis,
- the **p-value,** or **observed level of significance,** which equals the chance of getting a random sample that would make the test statistic at least as large as the actual value.

If the *p*-value is small then either the null hypothesis is false or we got a very unlikely sample.

Cautions Hypothesis tests and confidence intervals, like any tools, are designed to do particular jobs in particular situations. To use them wisely, you must pay attention to what they do *not* tell you, as well as what they do, and pay attention also to when the these tools may not be suitable. The methods of inference

- do *not* check the design of your study;
- do *not* tell you whether a result is important; and
- are *not* intended for data exploration.

9.2 CONFIDENCE INTERVALS

Constructing a confidence interval You can think of a confidence interval as a more formal version of "estimate plus or minus 2 *SE*s," based on the 68/95/99.7 rule for the normal distribution (Topic 3). If the sampling distribution is normal-shaped, then for 95% of all random samples, an interval of that form will contain the true value of the parameter. Using critical values from a table, we can also say that "estimate ± 1.65 *SE*s" gives a 90% confidence interval (90% of all random samples give an interval that contains the parameter), and "estimate ±2.58 *SE*s" gives a 99% confidence interval. In real applications, of course, we don't know the parameter value, and so we don't know, and can't tell, whether our interval scores a "hit" (covers the true value) or a "miss."

Choosing and comparing intervals Any confidence interval is based on a set of four quantities; any three of them determine the other one.

- *margin of error m:* narrower is better. It's much better to be able to say "The true value is within .05 of the estimate" than to have to say "The true value is somewhere within 5.0 of the true value."
- *confidence level C:* larger (closer to 1) is better. It's much better to be able to say, "For 99% of all random samples, my interval will cover the true value," than to have to say "In 50% of all random samples. . . ." Greater confidence comes at a price, however. For any given data set, the greater the confidence level you ask for, the wider the margin of error you must accept.

- *standard deviation of your response values:* smaller is better. Often, there's not much you can do about the SD of your response values. The SD is determined by the precision of your measuring process, or by the natural variation among the individuals in the population, or some combination of these. If you're lucky enough to have a small SD, then you can get narrow intervals with a high confidence level, but if you have a large SD, the only way to get narrow intervals with a high confidence level is to use a large sample size.
- *sample size n:* larger n gives narrower intervals and/or higher levels of confidence, but larger samples take more time and cost more.

You can use the formula $n = (z\sigma/m)^2$ to find out what sample size you need in order to get an interval with whatever margin or error and confidence level you want. (In the formula, z comes from a normal table and depends on the confidence level you want: For 90%, use $z = 1.65$; for 95%, use $z = 1.96$; for 99%, use $z = 2.58$.)

The meaning of 95% confidence It can be tempting to say that 95% confidence means that there is a 95% chance that the true value belongs to your interval. It would be very nice if we could say this and be correct, but, unfortunately, it's wrong. Ask yourself where the chance comes from. The chance is not in the unknown parameter (which doesn't move) but in the probability-based method of data production. *The chance is in the data, not in the parameter.* The population is fixed; what is random is the method of selecting the sample.

9.3 TESTS OF SIGNIFICANCE

The form of the test statistic For many situations, the test statistic has a particularly simple form: Use your sample to estimate the value of the parameter you want to know about; then ask, "How many SEs is it from my estimate to the null value?" Here's the logic. If the null hypothesis is true and your estimator is unbiased, then the *EV* of its sampling distribution is just the null value of the parameter, and the distance from your estimate to the *EV*, measured in *SEs*, is just the estimate converted to standard units.

$$\text{test statistic} = \frac{\text{estimate} - \text{null value}}{SE \text{ of the estimate}}$$

Two-tailed tests and one-tailed tests The right test to use depends on your **alternative hypothesis** H_A (or H_1), which in turn depends on the applied context: If it makes sense to consider parameter values on both sides of the null value, use a two-tailed test. If it makes sense to consider parameter values on just one side of the null value, use a one-tailed test. For example, to test for a *loss* in weight, an *improvement* in cure rate, or some other change that goes in just one direction, use a one-tailed test. To test whether the mean weight changed, or the cure rate for two procedures is different, use a two-tailed test.

Fixed-level testing For fixed-level testing, you choose a **significance level** α, such as $\alpha = .05$, in advance. Then you reject your null hypothesis if your observed p-value is less than or equal to α. With this form of testing, when we ask, "What will happen if I repeat this many times?" we can answer by giving the chances of rejecting H_0 when H_0 is true—a **Type I error**, or "false alarm" and of failing to reject H_0 when H_0 is false—a **Type II error**, or "miss." The chance of a false alarm is α, the level of the test. The chance of (correctly) rejecting H_0 when H_0 is false is called the **power** of the test.

Hypothesis tests and confidence intervals Both kinds of tools answer the question, "Which parameter values are reasonable, given the data?" A confidence interval gives a set of values considered reasonable; a significance test tells whether or not any given value can be considered reasonable. For most situations, the two procedures are equivalent, in that a 95% confidence interval is the set of parameter values not rejected by a two-tailed test at the 5% level.

Self-Testing Questions

1. *Confidence interval for a very simple estimation problem.* For this question, you are a statistician, and you agree to play the following game with me: I'm thinking of a number, the secret "true" value. Your job is to find a confidence interval for that true value. I won't tell you the number itself, but I will tell you the number I get by adding together my secret true value plus a chance error. To get my chance error, I'll draw at random from a box with four tickets.

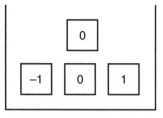

Suppose I tell you that I drew one chance error, that I added it to my true value to get an observed value, and that this observed value equals 19. Find a confidence interval for the true value.

➤ *Solution*

You might reason like this. There was a 50% chance (2 out of 4) that the chance error would be 0, which would make the observed value equal to the true value. So if you guess "True value equals 19," you'd be using a method that's right 50% of the time.

But there was a 25% chance that the ticket said −1, which would mean the true value was 20 (19 = 20 + [−1]). There's another 25% chance the ticket said +1, which would mean the true value was 18 (19 = 18 + [+1]). So if you guess "True value is between 18 and 20," you'd be using a method that's right 100% of the time.

In the language of confidence intervals, "18 to 20" is a 100% confidence interval for the true value, and "19 to 19" is a 50% "interval." (For this example, a 95% confidence interval doesn't exist.)

2. *Confidence interval when chance errors follow the normal curve.* The basic set-up is the same as in Question 1. I have a secret true value, and your job is to find a confidence interval. I'll draw one chance error at random, add it to the true value, and tell you the result. This time, however, I won't show you the box of chance errors. Instead, I'll tell you that there are 1000 tickets in the box, and that 950 of them are spread out between −10 and +10; the other 50 tickets are either less than −10 or bigger than +10. If my observed value (= true value + chance error) equals 19, find a 95% confidence interval for the true value.

➤ *Solution*

Even though you can't see the tickets, you know from what I've told you that 95% of them are between −10 and +10. Consider a few possible ways the draw from the box might turn out.

(i) If the chance error I draw is 10, then Obs = True + 10, and since Obs = 19, we have 19 = True + 10, and the true value must be 9.

(ii) If the chance error I draw is −10, then Obs = True − 10, and since Obs = 19, we have 19 = True − 10, and the true value must be 29.

(iii) If the chance error I draw is anywhere in between −10 and 10, then the true value must be somewhere in between 9 and 29.

Now think about the interval Obs ± 10 as a possible confidence interval. What is the chance this interval scores a "hit", that is, contains the true value? As long as the chance error I draw is one of the 950 tickets between −10 and 10, the distance from Obs to True will be at most 10, and Obs ± 10 will cover the true value. If, on the other hand, I draw one of the other 50 tickets, the distance from Obs to True will be more than 10, and Obs ± 10 will miss the true value. Thus each time you play this game with me, there's a 95% chance that your interval Obs ± 10 will cover the true value.

My actual observed value was 19. Does the interval from 9 to 29 contain the secret number I was thinking of? You'll never know, because I'm not going to tell you either the secret number or the chance error I got. That's how it is with confidence intervals in real life, too.

---

3. Now consider one last version of the question. This set-up is still basically the same: I tell you Obs (the sum of True + Chance error), and your job is to find a 95% confidence interval for the true value. This time, though, I won't even tell you how many tickets there are in the chance error box. All I'll tell you is that the numbers follow the normal curve, and have an average of 0 and a standard deviation of 5.

➤ *Solution*

I hope you recognize that this version of the problem leads to the same 95% confidence interval as before. If the chance errors follow the normal curve, then 95% of them are spread out within two SDs on either side of zero. Since the SD is 5, SD × 2 is 10, so 95% of the tickets are between −10 and 10. An interval of the form Obs ± 10 (or in general, Obs ± SD × 2) has a 95% chance of scoring a hit.

4. Suppose you want to construct a confidence interval based on a single observed value, and you know that the observed value equals an unknown true value plus a chance error drawn at random from the box below.

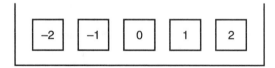

Suppose the observed value turns out to be 21.

(a) For this artificially simple example, you can be 100% sure, after observing the 21, that the true value is one of five numbers. List them.

(b) What is the confidence level (%) for the "interval" 21 to 21?

(c) What is the confidence level (%) for the interval 20 to 22?

Answer

(a) 19, 20, 21, 22, 23

(b) 20%

(c) 60%

5. Complete the sentence below in two different ways, each time choosing from the words in parentheses to fill in the blanks.

If the null hypothesis is in fact _____ (true/false) and the F-test _____ (rejects/fails to reject) it, the resulting error is called a _____ (miss/false alarm).

Answer

(a) true, rejects, false alarm

(b) false, fails to reject, miss

6. True or false.

(a) If the null hypothesis is true, two kinds of errors are possible: miss and false alarm.

(b) Hypothesis testing is conservative in that you declare the null hypothesis false unless there is strong evidence to the contrary.

(c) The critical values for testing at the 5% level have the property that if the null hypothesis is false, the chance of a miss is 5%.

(d) The critical value for testing at the 5% level is chosen so that 95% of all data sets for which the null hypothesis is true will have a value of the test statistic bigger than the critical value.

Answer

(a) F
(b) F
(c) F
(d) F

7. *Testing for ESP.* Identify each of the six parts of a statistical test (model, null hypothesis, test statistic, sampling distribution, observed outcome, conclusion) in this situation. Your friend claims he has ESP and can guess whether a card is red or black. To test his claim, you shuffle a deck of cards (half red, half black), draw one out at random, and ask him to guess. Then you shuffle, draw, and ask again, repeating this for a total of 20 guesses. You tentatively assume your friend doesn't really have ESP and is just guessing at random, but if he gets 15 or more right out of 20, you are prepared to conclude that there was more involved than pure guesswork. (Perhaps your friend is a skilled magician.) Just to complete the story, suppose that your friend gets 13 right.

Answer

Model: binomial. (**B**inary outcomes: right or wrong; **I**ndependent outcomes; **N** is fixed at 20; **S**ame probability p of a correct guess at each trial.)

Null hypothesis: $p = \frac{1}{2}$. (Your friend is just guessing.)

Test statistic: number of correct guesses = Y.

Sampling distribution: If H_0 is true, Y has the same distribution as the number of heads in 20 tosses of a fair coin.

Observed value: 13

Conclusion: The chance of 13 or more heads in 20 tosses of a fair coin is large enough (around 13%) that your friend could easily have gotten 13 right just by chance.

8. *Finding a sampling distribution.* Consider a very simple experiment, one with only one observed value. Assume the model is that the observed value equals true value plus chance error, with chance error determined by tossing a fair coin: Heads = +1, Tails = −1. The null hypothesis is that the true value equals 10.

(a) Suppose your test statistics is: (observed value − 10). Find the sampling distribution: assume the null hypothesis is true, list all possible values for the test statistic, and tell the chances.

(b) In part (a), what is your conclusion if the actual value of the test statistic is 1? 3?

(c) Now suppose our test statistic is (observed value − 10)^2. Find the sampling distribution.

Answer

(a) There are two possible values, 9 and 11, each with probability .5.

(b) If the test statistic equals 1 or 3, the null hypothesis can't be true.

(c) There is only one possible value, 1. (This would be a totally useless test statistic.)

9. *Finding a sampling distribution by simulation.* Consider a somewhat fancier version of the last experiment, this time with two observed values. The model is that each observed value equals (the same) true value plus chance error, and that to get each chance error, you toss three coins together and compute chance error = (# Heads − # Tails). Suppose, as in Question 8, your null hypothesis is that the true value equals 10. Take as your test statistic (average of observed values − 10)^2.

(a) Using three fair coins, create 20 data sets, compute the test statistic for each one, and summarize the results. (If you were to do this for 100 data sets instead of just 20, you'd have a decent approximate version of the sampling distribution.)

(b) Using your results from part (a) to judge, what would you conclude about the null hypothesis if a data set gave a value of the test statistic equal to 0? 16? 4? 9?

Answer

(a) Here are two possible samples:

Sample 1	Coin Tosses	Chance Error	Observation
Obs 1:	HHH	3 − 0 = 3	13
Obs 2:	THT	1 − 2 = −1	9
Average			11
Test statistic			1

Sample 2	Coin Tosses	Chance Error	Observation
Obs 1:	HHT	2 − 1 = 1	11
Obs 2:	HHH	3 − 0 = 3	13
Average			12
Test statistic			4

Here are the values of the test statistic from 20 simulations:

1, 4, 1, 4, 1, 1, 4, 0, 0, 1, 0, 1, 1, 1, 1, 1, 0, 1, 1, 1

And a summary:

Value of test statistic	0	1	4	9
Observed frequency	4/20	13/20	3/20	0/20
Theoretical frequency	10/32	15/32	6/32	1/32

(b) 0: not at all surprising if H_0 is true; do not reject H_0.
16: H_0 cannot possibly be true; reject H_0.
4: unusual if H_0 true, but not unusual enough to justify rejecting H_0.
9: highly unlikely ($p \approx .03$) if H_0 is true; reject H_0.

10. *Confidence intervals, Part 1.* Four box models for data are shown on p. 228. For each, first compute the *EV*. Then consider the following mechanism-and-summary: Draw once, construct a confidence interval of the form Draw ± 1, and record as your summary either "Hit" if the interval contains the *EV*, or "Miss" if it doesn't. Match each of the box models (a)–(d) on the left with the appropriate single-draw model (i)–(v) for the Hit/Miss summary on the right.

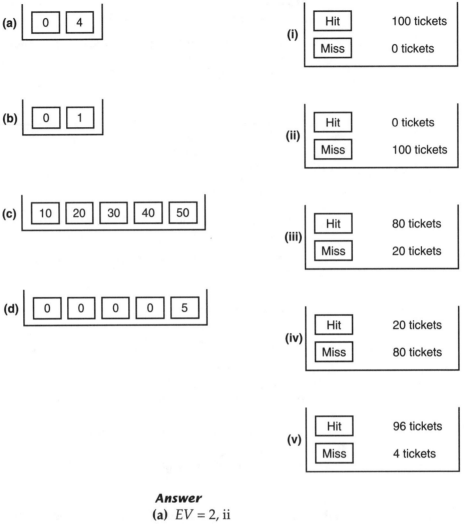

(a) | 0 | 4 |

(i) Hit — 100 tickets
Miss — 0 tickets

(b) | 0 | 1 |

(ii) Hit — 0 tickets
Miss — 100 tickets

(c) | 10 | 20 | 30 | 40 | 50 |

(iii) Hit — 80 tickets
Miss — 20 tickets

(d) | 0 | 0 | 0 | 0 | 5 |

(iv) Hit — 20 tickets
Miss — 80 tickets

(v) Hit — 96 tickets
Miss — 4 tickets

Answer
(a) $EV = 2$, ii
(b) $EV = 1/2$, i
(c) $EV = 30$, iv
(d) $EV = 1$, iii

11. *Confidence intervals, Part 2.* Use the same four box models for data as above, but this time use a different mechanism-and-summary: Draw twice, with replacement, compute the average of the two draws, construct the confidence interval as Ave ± 2, and record either "Hit" or "Miss" depending on whether your interval contains the EV for the box. Match each of the box models (a)–(d) on the left with the appropriate single-draw model (i)–(v).

Answer
(a) i
(b) i
(c) iv
(d) v

12. *Screening blood donors for hepatitis.* Although blood transfusions are usually safe, if you get a transfusion you do run a small risk of getting hepatitis from the donated blood. Fortunately, an effective test for screening donated blood can be based on the concentration of a particular enzyme, serum glutamic pyruvic transaminase (SGPT). People who are carriers of hepatitis tend to have higher levels of SGPT in their blood than non-carriers. It turns out that concentrations of SGPT have roughly a normal distribution, provided you measure in log (mg/100ml). Here's a model, in which Z has a standard normal distribution:

$$\text{Carriers: } \log \text{SGPT} = 1.55 + 0.13 \, Z$$
$$\text{Non-carriers: } \log \text{SGPT} = 1.25 + 0.12 \, Z$$

Suppose you run a blood bank, and you want to use blood concentrations of SGPT to screen out hepatitis carriers. Your idea is to choose a cut-off value: Anyone with log SGPT above that value you'll call a carrier, and not accept their blood; any one with log SGPT below that level, you'll call a non-carrier, and allow them to donate.

(a) *Two types of errors.* For each true condition, carrier and non-carrier, the test based on SGPT can either be right or wrong. Fill in the blanks to describe the two kinds of errors (choose from carrier, non-carrier, greater than, and less than): a Miss occurs if a _____ has log SGPT _____ than the cut-off, and gets called a _____; a False Alarm occurs if a _____ has a log SGPT _____ than the cut-off, and gets called a _____.

(b) For each of the possible cut-off values listed here, find the chance that (i) a carrier, and (ii) a non-carrier will be correctly classified. Cut-off values: 1.34, 1.40, 1.45.

(c) In your role as blood-banker, tell which kind of error from part (a) is more serious. Which of the three cut-off values in part (b) would you use if you ran the blood bank?

Answer

(a) carrier, less than, non-carrier
non-carrier, greater than, carrier

(b) Cut-off value of 1.34:

$$P(\text{Carrier} > 1.34) = P(1.55 + .13Z > 1.34)$$
$$= P(Z > (1.34 - 1.55)/.13)$$
$$= P(Z > -1.615)$$
$$= 0.95$$

$$P(\text{Non-carrier} < 1.34) = P(1.25 + .12Z < 1.34)$$
$$= P(Z < (1.34 - 1.25)/.12)$$
$$= P(Z < 0.75)$$
$$= 0.77$$

Here are the answers for the others:

	Probability of Correctly Classifying	
Cut-off Value	**Carrier**	**Non-carrier**
1.34	0.95	0.77
1.40	0.89	0.89
1.45	0.80	0.95

(c) It is much more serious to accept a carrier's blood ("miss") than to reject the blood of a non-carrier ("false alarm"). Of the three cut-off values, 1.34 gives the lowest chance of a miss.

13. Dr. Jones wants to know whether the drug Pravachol reduces cholesterol, so she sets up a randomized, double-blind experiment. After the data are collected she makes some calculations and does a hypothesis test to compare two means.

(a) In the context of this problem, explain what is meant by the statistical term "power."

(b) Suppose that Dr. Jones makes a mistake in her calculations and arrives at a p-value of .02, leading her to reject H_0 (with $\alpha = .05$), when the correct p-value is .09. While checking over the calculations of Dr. Jones, Dr. Smith notices the mistake and says "Dr. Jones has made a type I error." Is Dr. Smith's statement true? Why or why not?

Answer

(a) Power is the probability that we will determine, from the hypothesis test, that Pravachol works (by rejecting the null hypothesis) if, in fact, it does (i.e., if the null hypothesis is false).

(b) Dr. Smith is wrong. A type I error consists of rejecting a null hypothesis that is true. We do not know that the null hypothesis is true in this case (the *p*-value of .09 does not tell us that the null hypothesis is true), so we do not know whether a type I error was made.

14. Ben von Fischer (Oberlin College '96) took a large random sample of Oberlin students and asked them how many times they had bathed the previous week. Below are three confidence intervals for μ that were constructed from his data. Match the confidence levels with the intervals and explain your answer (briefly). (*Note:* You do not need to check the calculations, nor do you need any tables to answer this.)
(a) 90% **(i)** 5.84 ± .61 (5.23, 6.45)
(b) 95% **(ii)** 5.84 ± .80 (5.04, 6.64)
(c) 99% **(iii)** 5.84 ± .51 (5.33, 6.35)

Answer
(a) (iii) The 90% confidence interval will be the most narrow of the three.
(b) (i)
(c) (ii) The 99% confidence interval will be the most wide of the three.

15. Eight students from a statistics class attended a review session prior to the second exam. The average score among those 8 students was lower than the average for the 21 students who did not attend the review session. Suppose you want to use this information in a study of the effectiveness of review sessions.
(a) What kind of study is this: observational or experimental? Why?
(b) What kind(s) of sampling error(s) or bias(es) might be of concern here?
(c) (Hypothetical) You gave the data for the 8 who attended and for the 21 who did not attend to your friend George. George used the data to conduct a hypothesis test. Does a hypothesis test make sense? If so, what is H_0? If not, why not?

Answer

(a) This is an observational study, since the students chose for themselves which group to belong to.

(b) Selection bias and confounding are concerns: those who attend may need more help than those who do not attend. This could be why they have lower scores.

(c) A hypothesis test does not make sense. We don't have a random sample from a population; we have data on the entire class. You could think of the class as a sample of the students at the College, but it is not at all a random sample.

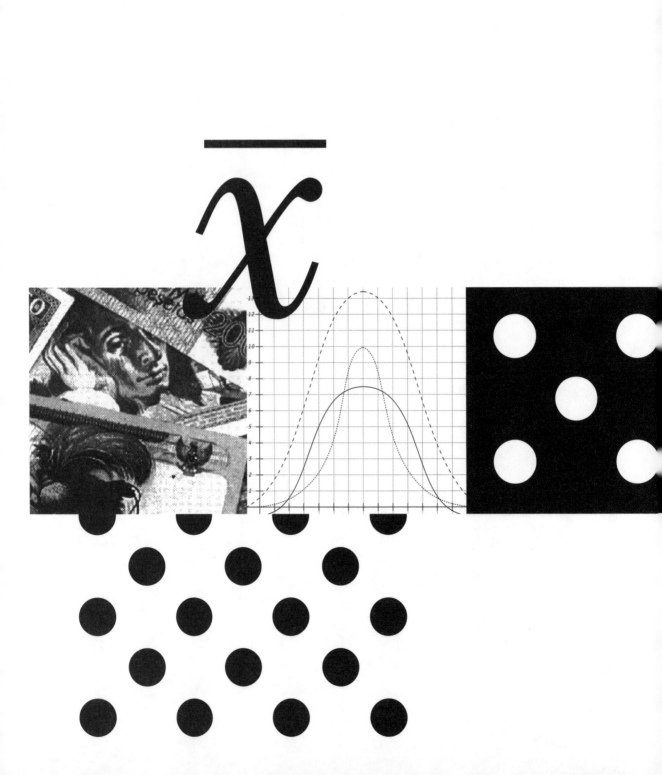

topic 10
Inference for Distributions

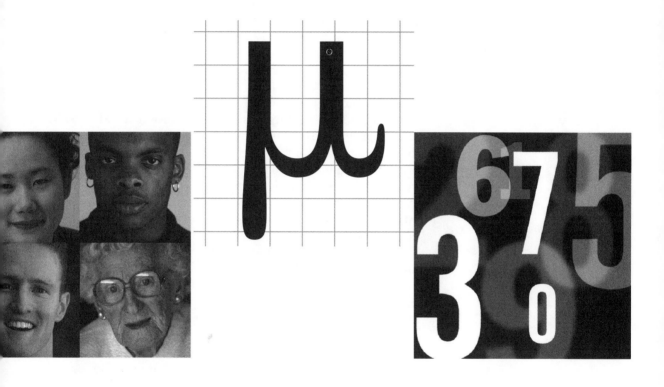

Summary _____

10.1 IN A NUTSHELL

Where does this topic fit in? The previous four topics (starting with Topic 6 on probability-based data production) have been building up the ideas you need in order to draw conclusions that go beyond the data you actually get to see. The topic just before this one is where all those ideas come together: A possible parameter value is judged reasonable if the resulting data summary is no more than about 2 *SE*s away from the parameter. Trial parameter values that make the data summary more than about 2 *SE*s away are judged unreasonable because they make the observed data too unlikely to have come from such a distribution. There are three closely related ways to put this logic into practice: confidence intervals, significance tests, and fixed-level testing.

This topic reviews these methods of inference for six of the simplest and most common situations in applied statistics. Most statistics books cover this material in two (or more) chapters, but they aren't consistent about how they divide things up, so the order here may not match your textbook. You may need to skip over some parts, and come back to others later on. The summaries are written in a way that makes this possible. As a result, later sections repeat some of the ideas from earlier sections, because not everyone will read the earlier ones first.

The deliberate repetition serves a second purpose as well: It helps you see how all the methods summarized here are just variations on a single set of ideas. I've grouped the methods together into one chapter because they all rely on the same logic, and they all follow the same pattern. First, they all ask, "How many *SE*s is it from an observed estimate to the true value?" Second, they use the sampling distribution of the estimate to ask, "How would that random distance behave if I were to repeat the process many times?" The answer, if samples are large enough: Expressed in standard units, the sampling distribution is approximately normal, with $EV = 0$ and $SD = 1$. This means that critical values for confidence intervals and p-values for hypothesis tests come from the standard normal distribution. (For smaller samples, the basic pattern has to be modified somewhat.)

Overview of the six situations You can think of the six sets of methods in a two-by-three table, organized into rows by the kind of response variable you have, and into columns by the kind of sample. If your response is binary (yes/no, success/fail-

ure, etc.), you should use a method in Row 1, which summarizes inference for proportions. If your response is quantitative,
you should use a method in Row 2, which summarizes inference for means. Column 1 corresponds to the simplest sampling
method, one simple random sample from a population. Column
2 is for paired data. It turns out that for both proportions and
for means, there is a way to use the one-sample method to analyze data for matched pairs. Finally, Column 3 is for two independent random samples.

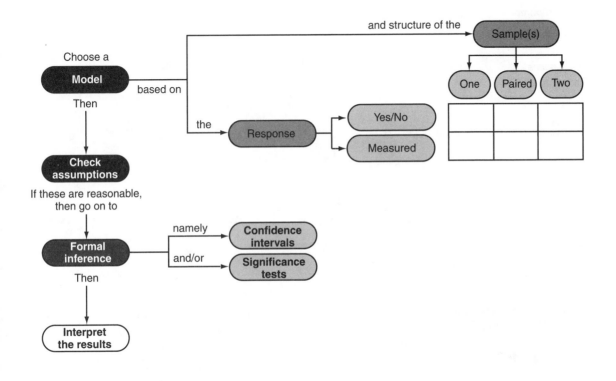

	Estimate	True Value	Standard Error	Interval	Test
			VALUE TO USE FOR:		
One proportion	$\hat{p}$	p	$\sqrt{p(1-p)}/\sqrt{n}$	$\hat{p}$	p_0
One mean	$\bar{y}$	μ	$\sigma/\sqrt{n}$	s	s
Two proportions	$\hat{p}_1 - \hat{p}_2$	$p_1 - p_2$	$\sqrt{\dfrac{p_1(1-p_1)}{n_1} + \dfrac{p_2(1-p_2)}{n_2}}$	$\hat{p}_1, \hat{p}_2$	$\hat{p}$
Two means	$\bar{y}_1 - \bar{y}_2$	$\mu_1 - \mu_2$	$\sqrt{\dfrac{\sigma_1^2}{n_1} + \dfrac{\sigma_2^2}{n_2}}$	s_1^2, s_2^2	s_1^2, s_2^2

Test statistic = (Est − True)$/SE$ =

Confidence interval: Est ± (Critical value)SE =

10.2 CATEGORICAL RESPONSE: ONE PROPORTION

Example. Here's an activity you can try yourself. Get a penny (earlier dates work better) and place it on edge on a table, holding it with your index finger. Adjust the penny so that Lincoln's head is facing you, right side up. Then flick the penny with one finger of the other hand, and record whether it lands with Heads or Tails facing up. Repeat this 25 times to get your sample. (It turns out that when you spin a penny, Heads and Tails are not equally likely.)

As you read through the summary that follows, you can think back to this example as a way of making the description more concrete. For data, you can do the activity yourself, or else use these results: 8 heads in 25 spins.

The model Your data must come from a binomial process, with the four properties from Topic 7:

- **B** inary outcomes;
- **I** ndependent outcomes;
- **N** umber n of trials is fixed in advance;
- **S** ame chance p of success on each trial.

Perhaps the most common situation is a simple random sample taken from a population of binary outcomes. This is the main requirement, but in addition, there are two more technical requirements that allow you to use the binomial SE and the normal approximation. First, if you are sampling without replacement, which is usually the case in real applications, the formula for the standard error requires that your sample be no larger than 10% of the population.[1] Second, in order to use the normal approximation, you need a sample large enough that the expected number of outcomes of each kind is at least 10: $np \geq 10$ and $n(1 - p) \geq 10$.

Confidence interval: estimate ± margin of error

- Take as your **estimate** the observed fraction of successes: $\hat{p} =$ (# successes)$/n$.
- The **margin of error**, as always, equals the critical value from a table, times the standard error. (a) If the conditions listed above are satisfied, then your critical value z comes from a normal table, and depends on the level of confidence you want. For 90% confidence, use $z = 1.65$; for 95%, use $z = 1.96$; and for 99%, use $z = 2.58$. (b) For a single binomial sample, $SE = \sqrt{\dfrac{\hat{p}(1 - \hat{p})}{n}}$.
- The **interpretation** is that, if you were to repeat the process many times, then for 95% of all samples (or 90% or 99%), the interval *estimate ± margin of error* will cover the true value.

---

[1] Otherwise, you should use $SE = $ (binomial SE) × (correction factor) = $\sqrt{\dfrac{p(1 - p)}{n}} \times \sqrt{\dfrac{N - n}{N - 1}}$. In the formula for the correction factor, N, is the size of the population, and n is the size of the sample. Thus the correction factor is approximately the square root of the unsampled fraction of the population. Notice that if your sample is a tiny fraction of the population, the correction factor is very close to 1; if you sample the whole population, the correction factor is 0, which tells you there is no variability in your estimate: You know all there is to know about the population.

(Remember, the chance is in the sampling process, not in the parameter.)

Testing hypotheses

- Your **null hypothesis** is that the true proportion p in the population equals a particular value p_0. Should you use a one-sided or two-sided **alternative**? You should decide this based on what you know about the biology, psychology, economics, etc., of the applied context. Before you see the data, can you rule out one or the other of $p < p_0$ or $p > p_0$ as unreasonable, or uninteresting, or irrelevant? If so, you'll want to exclude that possibility from consideration by using a one-sided test. Otherwise, use a two-sided test.
- The **test statistic** tells how many SEs it is from the observed estimate $\hat{p}$ to the null value p_0. (To compute the standard error, assume the null hypothesis is true:

$$SE = \sqrt{\frac{p_0(1 - p_0)}{n}}.$$ Large distances (more than 2 SEs or so) are

unlikely if the null hypothesis is true, and so they provide evidence against H_0. Distances that are not large are about what you would expect if H_0 were true, and so give no reason to be suspicious of H_0. To quantify this logic we use the sampling distribution to compute a p-value (or observed significance level).
- The **p-value** is based on the question, "If I were to repeat this (gather-the-data-and-compute-the-test statistic) a large number of times, what is the chance I'd get a value of the test statistic at least as large as the value from my actual data?" Here, the meaning of "at least as large" depends on the tails of your test. For a left-tailed test, use the area (probability) to the left of your test statistic; for a right-tailed test, use the area to the right. For any given test, the left-tailed and right-tailed p-values add to 1, and the two-tailed p-value is twice the smaller of the one-tailed p-values. For fixed level testing, you reject your null hypothesis if the p-value is less than α, the level of your test.

10.3 PROPORTIONS IN MATCHED PAIRS (MCNEMAR'S TEST)

Example. The first U.S. study of links between using oral contraceptives and thromboembolic disease (blood clots) appeared in 1969. Hospital records were used to identify 175 pairs of women, one in each pair with the disease and one without. Within each pair, the women were matched by age, marital status, residence, race, number of prior pregnancies, and whether they had stayed in a private room, semi-private room, or ward. The response here is whether (yes/no) a woman had used oral contraceptives. Among the 175 women with thromboembolic disease, 67, or 38%, had used oral contraceptives, whereas among the 175 controls, only 23, or 13%, had.

Oral Contraceptives and Thromboembolic Disease

There were 175 matched pairs. Within each pair, one woman (the "case") had the disease, the other (the "control") did not. The table summarizes their use of oral contraceptives.

Oral Contraceptives Used by:	Number of Pairs
Both women in the pair	10
Only the case	57
Only the control	13
Neither of the women	95
Total	175

We have a sample of matched pairs with a binary response, and we want to test whether the difference between the two sample proportions could be due to chance. **McNemar's test** gives us a way to turn our data into a single binomial sample.

McNemar's test Sort the data into four kinds of pairs, as in the table. Notice that pairs for which the outcomes are the same—Yes, Yes (both used contraceptives) and No, No (neither did)—give us no information about whether the two proportions are equal. Ignore these pairs. That leaves only two kinds of pairs—Yes, No (only the woman with blood clots used oral contraceptives), and No, Yes (only the control did). We can treat these 70 pairs (57 Yes, No and 13 No, Yes) as a single binomial sample, and test as our null hypothesis that the two kinds of pairs are equally likely. It turns out that the observed proportion $57/70 = .814$ is more than 5 SEs from the null value of .5: We reject H_0.

10.4 COMPARING PROPORTIONS IN TWO INDEPENDENT SAMPLES

Example. The video segment on the CD tells about contaminated drinking water in Woburn, Massachusetts. You can use that example as a way of making the summary that follows more concrete. There were two samples of babies, one from the east side of town where the residents were getting contaminated water, and a second sample from the west side, whose residents got their water from other, uncontaminated wells. The sample from the east showed 16 birth defects out of 414 births. The sample from the west showed only 3 birth defects, although the sample size was quite a bit smaller, too, with only 228 births.

The model For inferences to be valid, the sampling distribution you use must match the kind of data you have. For the method described here, your data must come from two independent binomial processes, each characterized by the usual four properties from Topic 7 (**B**inary outcomes, **I**ndependent outcomes, **N**umber n of trials is fixed in advance, **S**ame chance p of success on each trial). For example, you might have simple random samples taken from two different populations with binary outcomes. This is the main requirement, but in addition, just as with a single binomial sample, there are two technical requirements that allow you to use the binomial SE and the normal approximation. These are exactly the same as before, and must be true of both samples. First, if you are sampling without replacement, the formula for the standard error requires that your sample be no larger than 10% of the population.[2] Second, in order to use the normal approximation, your two samples must be large enough that the expected number of outcomes of each kind is at least 5: $n_1 p_1 \geq 5$, $n_1(1 - p_1) \geq 5$, $n_2 p_2 \geq 5$, and $n_2(1 - p_2) \geq 5$.[3]

Confidence interval: estimate ± margin of error

- Take as your **estimate** the difference in the observed fraction of successes: $p_1 - p_2 =$ (first proportion minus second proportion).
- The **margin of error,** as always, equals the critical value from a table times the standard error. (a) If the conditions listed above are satisfied, then your critical value z comes from a normal table, and it depends on the level of confidence you want. For 90% confidence, use $z = 1.65$; for 95%, use $z = 1.96$; and for 99%, use $z = 2.58$. (b) For the SE, compute separate SEs for the two sample proportions, and then "add them like the Pythagorean theorem:" Square each SE, add, then take the square root. This gives
$$\sqrt{\frac{\hat{p}_1(1 - \hat{p}_1)}{n_1} + \frac{\hat{p}_2(1 - \hat{p}_2)}{n_1}}.$$

---

[2] Otherwise, just as with one sample, you should use the corrected $SE =$ (binomial SE) × (correction factor) $= \sqrt{\dfrac{p(1 - p)}{n}} \times \sqrt{\dfrac{N - n}{N - 1}}$.

[3] If your samples are too small to meet this condition, then you should use Fisher's exact test, which is described in books on statistical methods and is available in many software packages.

- The **interpretation** is that if you were to repeat the process many times, then for 95% of all samples (or 90% or 99%), the interval *estimate ± margin of error* will cover the difference $p_1 - p_2$ between the true population proportions. (Remember, the chance is in the sampling process, not in the parameters.)

Testing hypotheses

- Your **null hypothesis** is that the two proportions are equal (H_0: $p_1 = p_2$). Should you use a one-sided or two-sided alternative? As with a single sample, you should decide this based on what you know about the applied context. Before you see the data, can you rule out one or the other of $p_1 < p_2$ or $p_1 > p_2$ as unreasonable, or uninteresting, or irrelevant? If so, you'll want to exclude that possibility from consideration by using a one-sided test. Otherwise, use a two-sided test.
- The **test statistic** tells how many *SEs* it is from the observed estimate $\hat{p}_1 - \hat{p}_2$ to the null value of 0 ($p_1 = p_2$). To measure this distance, we use a standard error computed assuming the null hypothesis is true, that is, $p_1 = p_2$. To estimate this common value, combine the two samples and treat them as a single sample. Then the SE for our test statistic is $\sqrt{\dfrac{\hat{p}(1 - \hat{p})}{n_1 + n_2}}$.

  (*Careful:* This is a different *SE* from the one you use for a confidence interval!) As before, large distances (more than 2 *SEs* or so) are unlikely if the null hypothesis is true, and they provide evidence against H_0. Distances that are not large are about what you would expect if H_0 were true, and so give no reason to be suspicious of H_0. To quantify this logic we use the sampling distribution to compute a *p*-value (or observed significance level).
- The **p-value** is based on the question, "If I were to repeat this (gather-the-data-and-compute-the-test-statistic) a large number of times, what is the chance I'd get a value of the test statistic at least as large as the value from my actual data?" Here, as with one sample, the meaning of "at least as large" depends on the tails of your test. For a left-tailed test, use the area (probability) to the left of your test statistic; for a right-tailed test, use the area to the right. For any given test, the left-tailed and right-tailed *p*-values add to 1, and the two-tailed *p*-value is twice the smaller of the one-tailed *p*-values. For fixed level testing, you reject your null hypothesis if the *p*-value is less than α, the level of your test.

10.5 QUANTITATIVE RESPONSE: ONE MEAN

Example. The CD contains a video segment to illustrate this kind of situation: A laboratory that supplies chemical reference materials needs to be able to give a confidence interval for the concentration of PCBs (polychlorinated biphenols, a dangerous pollutant) in their reference vials of PCBs. They take a random sample of ten vials, and measure the concentrations, in parts per billion: 289, 287, 293, 290, 287, 289, 287, 289, 291, 289. The sample mean is 289.1, and the sample standard deviation is 1.912.

The model For your inferences to be valid, your data must come from a process that matches the sampling distribution you use. To use the method summarized here, your data must meet two requirements. First, the chance model: Your data must come from a simple random sample, a completely randomized experiment, or some other chance process that generates independent outcomes from the same distribution. Second, the sample size and shape. For small samples, your data should come from a nearly normal-shaped distribution. For larger samples, your data can come from a distribution that is moderately skewed. The larger your sample, the less the shape of the distribution tends to matter because methods here, which are based on the *t*-distribution, are fairly robust.

Confidence interval: estimate ± margin of error

- Take as your **estimate** the sample mean $\bar{y}$.
- The **margin of error**, as always, equals the critical value from a table times the standard error. (a) If the conditions listed above are satisfied, then your critical value t comes from a table of the *t*-distribution, with degrees of freedom equal to $n - 1$, and, as usual, depends on the level of confidence you want. (b) Your (estimated) standard error is $SE = \frac{s}{\sqrt{n}}$. (The reason for using a *t*-distribution instead of the normal for the critical value is to adjust for the variability in the sample standard deviation s.)
- The **interpretation** is that, if you were to repeat the process many times, then for 95% of all samples (or 90%, or 99%), the interval *estimate ± margin of error* will cover the true value. (Remember, the chance is in the sampling process, not in the parameter.)

Testing hypotheses

- Your **null hypothesis** is that the true population mean is equal to some particular value μ_0. Should you use a one-sided or two-sided **alternative**? You should decide this based on what you know about the biology, psychology, economics, etc., of the applied context. Before you see the data, can you rule out one or the other of $\mu < \mu_0$ or $\mu > \mu_0$ as unreasonable, or uninteresting, or irrelevant? If so, you'll want to exclude that possibility from consideration by using a one-sided test. Otherwise, use a two-sided test.
- The **test statistic** tells how many *SE*s it is from the observed estimate $\bar{y}$ to the null value μ_0. (The *SE* here is the same as for the confidence interval.) Large distances (more than 2 SEs or so) are unlikely if the null hypothesis is true, and they provide evidence against H_0. Distances that are not large are about what you would expect if H_0 were true, and so give no reason to be suspicious of H_0. To quantify this logic we use the sampling distribution to compute a *p*-value (or observed significance level).
- The *p*-**value** is based on the question, "If I were to repeat this (gather-the-data-and-compute-the-test statistic) a large number of times, what is the chance I'd get a value of the test statistic at least as large as the value from my actual data?" Here, the meaning of "at least as large" depends on the tails of your test. For a left-tailed test, use the area (probability) to the left of your test statistic; for a right-tailed test, use the area to the right. For any given test, the left-tailed and right-tailed *p*-values add to 1, and the two-tailed *p*-value is twice the smaller of the one-tailed *p*-values. For fixed level testing, you reject your null hypothesis if the *p*-value is less than α, the level of your test.

10.6 COMPARING MEANS FOR MATCHED PAIRS

Example. On the CD, there is a video segment that shows a matched pairs design. Each pair corresponds to a person who rated the sweetness of a soft drink made with Nutrasweet both before and after storage at a high temperature. The goal of this part of the Nutrasweet taste test was to see whether the soft drink lost sweetness during storage. Each person provided a pair of response values for before and after storage. To test whether the mean sweetness after storage is the same as the mean sweetness before, we can look at the difference, (after − before) and test that the true mean difference is 0. This way of looking at the problem allows you to apply the one-sample *t*-test for the mean of one sample. Here are the differences for the ten raters: 2.0, 0.4, 0.7, 2.0, −0.4, 2.2, −1.3, 1.2, 1.1, 2.3. The sample mean difference is 1.02, and the sample standard deviation of the ten differences is 1.196.

10.7 COMPARING MEANS FROM TWO INDEPENDENT SAMPLES

Example. The CD shows a laboratory comparing Ultracell, a new kind of foam for cushions, to the standard foam used in cushions. One of the tests compares the two foams for springiness by dropping a ball bearing on the foam cushions and measuring the height of the bounce as a percentage of the initial height. Here, each bounce is a case, and we have two samples of bounces: one sample using Ultracell, and the other sample using the standard foam. For the Ultracell sample of 13 bounces, the mean bounce was 65% of the initial height, with sample standard deviation of 2.5. For the sample of 10 bounces on standard foam, the mean height was 50% of the initial height, and the sample standard deviation was also 2.5.

The model For your inferences to be valid, your data must come from a process that matches the sampling distribution you use. To use the method summarized here, your data must meet two requirements. First, the chance model: Your data must come from two simple random samples from different populations, from a completely randomized experiment, or from some other chance process that generates independent outcomes from two fixed distributions. Second, the size and shape: For small samples (combined size less than 15), your data should come from a nearly normal-shaped distribution. For larger samples, your data can come from a distribution that is moderately skewed. The larger your sample, the less the shape of the distribution tends to matter because methods here, which are based on the *t*-distribution, are fairly robust.

Confidence interval: estimate ± margin of error

- Take as your **estimate** the difference between your two sample means: $\bar{y}_1 - \bar{y}_2$.
- The **margin of error**, as always, equals the critical value from a table times the standard error. (a) If the conditions listed above are satisfied, then your critical value *t* comes from a table of the *t*-distribution, with degrees of freedom that depend on how you estimate the *SE*, and also, as usual, on the

level of confidence you want. (b) There are two ways to estimate the SE, summarized in the table. The two-sample estimate is generally the better choice because it does not require
you to assume that your two populations have the same standard deviation, an assumption that is not easy to check and is
not true of many data sets. Although the resulting sampling
distribution when you use the two-sample estimate is only
approximately a t-distribution, the approximation is generally quite good. The other way to estimate the SE, using a
pooled estimate of the SD, is much more specialized because
of the extra assumption, and a test or interval using this estimate may give misleading results when your data don't fit the
assumption.

- The **interpretation** of your interval is the usual. If you were to
repeat the process many times, then for 95% of all samples (or
90% or 99%), the interval *estimate ± margin of error* will cover
the true value. (*Remember, the chance is in the sampling process,
not in the parameter.*)

Standard Errors and Degrees of Freedom

Method	Standard Error	Degrees of Freedom
Two-sample: does not require $\sigma_1^2 = \sigma_2^2$	$\sqrt{\dfrac{s_1^2}{n_1} + \dfrac{s_2^2}{n_2}}$	Smaller of n_1-1, n_2-1*
Pooled estimate: requires $\sigma_1^2 = \sigma_2^2$	$\sqrt{\dfrac{(n_1 - 1)s_1^2 + (n_2 - 1)s_2^2}{n_1 + n_2 - 2}\left[\dfrac{1}{n_1} + \dfrac{1}{n_2}\right]}$	$n_1 + n_2 - 2$

Testing hypotheses
- Your **null hypothesis** is that the true population means are
the same: $\mu_1 = \mu_2$. To decide whether to use a one- or two-
sided **alternative**, rely, as usual, on the applied context. Can
you rule out one or the other of $\mu_1 < \mu_2$ or $\mu_1 > \mu_2$ as unreasonable, or uninteresting, or irrelevant? If so, you'll want to

*Computer software uses $df = \dfrac{\left[\dfrac{s_1^2}{n_1} + \dfrac{s_2^2}{n_2}\right]^2}{\dfrac{1}{(n_1 - 1)}\left[\dfrac{s_1^2}{n_1}\right]^2 + \dfrac{1}{(n_2 - 1)}\left[\dfrac{s_2^2}{n_2}\right]^2}$

exclude that possibility from consideration by using a one-sided test. Otherwise, use a two-sided test.

- The **test statistic** tells how many *SE*s it is from the observed estimate $\bar{y}_1 - \bar{y}_2$ to the null value of 0. (The *SE* here is the same as for the confidence interval.) Large distances (more than 2 *SE*s or so) are unlikely if the null hypothesis is true, and they provide evidence against H_0. Distances that are not large are about what you would expect if H_0 were true, and so give no reason to be suspicious of H_0. To quantify this logic we use the sampling distribution to compute a *p*-value (or observed significance level).

- The ***p*-value** is based on the question, "If I were to repeat this (gather-the-data-and-compute-the-test statistic) a large number of times, what is the chance I'd get a value of the test statistic at least as large as the value from my actual data?" Here, the meaning of "at least as large" depends on the tails of your test. For a left-tailed test, use the area (probability) to the left of your test statistic; for a right-tailed test, use the area to the right. For any given test, the left-tailed and right-tailed *p*-values add to 1, and the two-tailed *p*-value is twice the smaller of the one-tailed *p*-values. For fixed level testing, you reject your null hypothesis if the *p*-value is less than α, the level of your test.

10.8 INFERENCES FOR STANDARD DEVIATIONS

Inferences about standard deviations depend on the shape of the distributions. The most common test, based on the *F*-distribution, assumes normal shapes, and is very sensitive to departures from normality. It is not recommended.

Self-Testing Questions _____

1. Riton is a protease inhibitor which may reduce by half the death rate from advanced AIDS and the serious complications due to the disease. In a *New York Times* article (2/2/96), 26 of a sample of 543 who received riton died of AIDS, while 46 of the 547 who received a placebo died of AIDS.
   (a) Is it appropriate to use a normal model with these data?
   (b) What is the null hypothesis?
   (c) What is the alternative hypothesis?

(d) If H_0 is true, find the pooled estimate $\hat{p}$.
(e) What is the *SE*?
(f) What is the difference of the proportions?
(g) Is there a significant difference at the .05 level?

Answer
(a) yes, since all expected counts are greater than 10.
(b) *P*(death on riton) = *P*(death on placebo)
(c) *P*(death on riton) < *P*(death on placebo)
(d) .066
(e) .015
(f) .0362
(g) Yes, *Z* = −2.41, *p*-value = .008

2. Use the data from Question 1 to find a 90% confidence interval for the difference in population proportions.

   Answer (.012, .061)

3. Interpret the confidence interval from Question 2.

   Answer

   We are 95% confident that the probability of dying when on riton is between 1.2% and 6.1% lower than the probability of dying when on the placebo.

4. A survey of 62 students at Grinnell College found that 8 out of 24 first-year students preferred Grinnell water over other water choices, while 29 out of 38 second-year students showed the same preference.
   (a) In doing a statistical analysis of this data, what is the null hypothesis?
   (b) The purpose of the test is to find out if student attitudes change in a positive way toward Grinnell's water as they stayed in Grinnell longer. What is the alternative hypothesis in this case?
   (c) What is the difference in sample proportions?

(d) Is the difference significant at the .1 level? The .05 level? The .01 level?

Answer
(a) p(first years) $= p$(second years)
(b) p(first years) $< p$(second years)
(c) .43
(d) yes in all cases, since $z = 3.36$.

5. Use the data from Question 4 to construct a 95% confidence interval for the difference in population proportions.

Answer (.198, .662)

6. Are health care workers who wear rings more susceptible to spreading bacteria? Thirty-two students served as subjects for a study of this question by Jacobsen et al. (Nursing research, 1985). Bacteria counts were made on each student's hands after careful washing. On one day this was done when the students were wearing rings; on another day, the students were not wearing rings. For each student, the difference—count without rings minus count with rings—was obtained. These 32 differences had an average of 842 and a standard deviation of 4215. At the 10% level, do these data provide statistically significant evidence that wearing rings leads to higher bacteria counts on average?

➤ **Solution**

The t-ratio of $842/(4215/\sqrt{32}) = 1.13$ has 31 degrees of freedom. The one-sided p-value is greater than 10%. Thus, the evidence is not significant.

7. Students wondered if right-handers could do more arm curls with their right arms than with their left. Eighteen right-handed subjects were recruited to perform an experiment with 10-pound dumbbells. The number of repetitions to exhaustion for a set exercise (floor to side to chest to side to floor) were measured for each subject on each arm. The beginning arm was randomized between right and left. The subjects averaged 43.33 arm curls with their right arms and

37.11 with the left arms. What is the best way to analyze this data?

> **Solution**

Since the measurements are made for right and left arms on the same subjects, these data should be analyzed as a matched-pairs design.

8.

(a) A survey of 200 college students found that only 50 agreed with the President's foreign policy. What is the *SE* of the sample proportion?

(b) What is the length of the confidence interval at the 95% level?

Answer

(a) .0306

(b) $2 \times 1.96 \times .0306 = .12$

9. Consider the following data based on the number of mosquito bites received as a test of a new insect repellent.

Person	Without	With	Difference
1	27	33	−6
2	9	9	0
3	33	21	12
4	33	15	18
5	4	6	−2
6	22	16	6
7	21	19	2
8	33	15	18
9	20	10	10

(a) What is the *t*-statistic for this set of paired data?

(b) Is the difference significant at the .05 level?

Answer

(a) $t = \dfrac{6.44 - 0}{8.65/\sqrt{9}} = 2.23$

(b) No

10. Use the data from Question 9 to construct a 95% confidence interval for the mean improvement in the entire population.

Answer

$$6.44 \pm 2.306 \times \frac{8.65}{\sqrt{9}} \text{ or } (-.21, 13.09)$$

11. Two roommates are trying to decide who can better predict the length of their phone calls. As a result, before each call they make, they make an estimate of how long that call will take and compare that estimate to the actual length of the call (no peeking at the clock is allowed). Is this paired data?

➤ **Solution**

No. While the prediction and actual length can be paired, the comparison is between the unpaired performance (as measured in that difference) between the two roommates.

12. Eleven runners were randomly chosen to investigate body dehydration. Plasma concentrations (in pmol/liter) were measured before and after the Tyneside Great North Run. Larger plasma concentrations imply greater body dehydration.
(a) What method should be used to analyze these data?
(b) What is the most appropriate alternative hypothesis?

Answer

(a) Since the prerace and postrace measurements were made on the same runners, these data should be analyzed as a matched-pairs design.
(b) The population postrace mean exceeds the population prerace mean.

13. The distribution of systolic blood pressure (SBP) in the general population is approximately normally distributed, with a mean of 130mm Hg and a standard deviation 20mm Hg. In a random sample of 25 glaucoma patients, the mean SBP was 136mm Hg and the standard deviation was 15mm Hg. Do these data provide statistically significant evidence that, on average, systolic blood pressure is higher for glaucoma patients? Use a 5% significance level and a one-sided alternative.

➤ *Solution*

Since the t-ratio = 2.0, we have statistically significant evidence that mean blood pressure levels are higher for glaucoma patients.

---

14. Use the data in Question 13 to construct a 95% confidence interval for the mean SBP of glaucoma patients.

Answer

$$136 \pm 2.064 \times \frac{15}{\sqrt{25}} \text{ or } (129.8, 142.2)$$

15. Interpret the confidence interval from Question 14.

Answer

We are 95% confident that the average SBP for all glaucoma patients is between 129.8 and 142.2.

16. The confidence interval for Question 14 includes 130. Why doesn't this contradict the answer to Question 13?

➤ *Solution*

Question 13 used a one-sided H_A. The CI in Question 14 corresponds to a test with a two-sided H_A (for which we would not reject H_0).

---

17. A t-distribution with 14 degrees of freedom has an 80th percentile of .8681. Is it true that the 80th percentile of a t-distribution with 16 degrees of freedom will be a number larger than .8681?

➤ *Solution*

False, since the t-distribution with 16 degrees of freedom is less spread out, its 80th percentile will be smaller than the 80th percentile of the t-distribution with 14 degrees of freedom.

---

18. A random sample of 17 lengths of a dimension on an integrated circuit produced a sample mean of 15.3 microns and a sample standard deviation of 1.6 microns. A confidence interval for the true process mean was calculated from these data, using the usual method based on the t-distribution.

However, it was then discovered that the 17 data values contained two outliers. One was usually large and the other was unusually small. (Both were caused by inadvertently interchanging digits when the data were entered into the computer.) What effect will this error have on the width of the resulting confidence interval?

➤ **Solution**

The outliers will make the confidence interval too wide, since the sample standard deviation will be too large.

---

19. We have observed a random sample of 9 values from a stable normal process and obtained a sample mean of 100.5 and sample standard deviation of .23. We need to estimate the mean of the process. What is the length of a 95% confidence interval for the process mean?

Answer

$$2 \times 2.306 \times \frac{.23}{\sqrt{9}} = .354$$

20. Suppose that Y1, Y2, . . . ,Y7 is a random sample of size 7 from a normal population with standard deviation σ. Let $\overline{Y}$ denote the sample mean and S the sample standard devia–tion. What, then, is the distribution of the quantity $\dfrac{\overline{Y} - \mu}{s\sqrt{7}}$?

Answer

A *t*-distribution with 6 degrees of freedom

21. The winning times for the Boston Marathons for the years 1968–1978 have mean = 135 and SD = 4; for 1979–1989, the mean is 130 and SD is 1.9. Can we conduct a hypothesis test to compare the two means?

➤ **Solution**

No. We have all of the data, not random samples from large popula-tions. We know the averages, so there is nothing to test.

---

22. Eight students measured their hand strength by squeezing a bathroom scale, first with their right hand and then with their left hand. On average, the right hand strength was 3 pounds more than the left hand strength. The SD of the differences was 3.4. Use these data to construct a 95% confidence interval.

Answer

$$3 \pm 2.365 \times \frac{3.4}{\sqrt{8}} \text{ or } (.16, 5.84)$$

23. Interpret the CI from Question 22.

Answer

We are 95% confident that right hand strength exceeds left hand strength by between .16 and 5.84 pounds, on average.

24. Use the data from Question 22 to conduct a test of H_0: $\mu = 0$ versus H_A: $\mu \neq 0$. Use $\alpha = .05$.

➤ *Solution*

$$t = \frac{3 - 0}{3.4/\sqrt{8}} = 2.496, \text{ so we reject } H_0.$$

25. What assumptions are necessary for the test in Question 24 to be valid?

➤ *Solution*

We must assume that we have a random sample and that the distribution of right hand strength minus left hand strength follows a normal curve.

26. The average SAT verbal score for a group of 100 students at a university was 520, with a standard deviation of 115. What is the standard error of the mean?

Answer

$$\frac{115}{\sqrt{100}} = 11.5$$

27. Use the data in Question 26 to construct a 95% confidence interval for the mean.

Answer

$$520 \pm 1.96 \times \frac{115}{\sqrt{100}} \text{ or } (497.5, 542.5)$$

28. The national average SAT verbal score was 505. Do the data in Question 26 provide evidence, at the .05 level of significance, to indicate that this university differs from the rest of the nation?

➤ ***Solution***

No. $z = \dfrac{520 - 505}{115/\sqrt{100}} = 1.30$. The p-value is .195.

29. What assumptions are necessary for the test in Question 28 to be valid?

➤ ***Solution***

We need a random sample. (Note that the distribution of SAT scores does not have to be normal since it is large.)

30. Kiki Bennett (Oberlin College '96) collected data on the GPAs of 14 students who say that the Mandarin is their favorite Oberlin restaurant. The sample average of these 14 GPAs was 3.16 and the sample SD was .31. We wish to use this information to conduct a test of the null hypothesis that the population average for Mandarin-preferring students is 2.93 (which is the overall average GPA of all Oberlin students). Use a two-sided alternative.
(a) State H_0 and H_A in symbols.
(b) Conduct the test. Provide all the steps, including degrees of freedom and bounds on the p-value. If $\alpha = .05$, do you reject H_0? Why or why not?
(c) What assumptions are necessary for your test in part (b) to be valid?

Answer
(a) H_0: $\mu_1 = 2.93$
    H_A: $\mu_1 \neq 2.93$

(b) The standard error is $\dfrac{.31}{\sqrt{14}} = .083$. The test statistic is $t = \dfrac{3.16 - 2.93}{.083} = 2.77$. The t statistic has 13 degrees of freedom, so the p-value is between .01 and .02. (More precisely, the p-value is .016). We reject H_0 because the p-value is less than α.

(c) We must assume that the data are a random sample from a normal population.

31. Xinchen Lou (Oberlin College '97) collected data on the number of hours of sleep that students get in a night. The sample average of 29 seniors as 6.90, with a sample of SD of 1.92. For 30 first-year students the sample average was 6.08 and the sample SD was 1.73. The standard error for the difference in the sample averages, $\sqrt{\dfrac{s_1^2}{n_1} + \dfrac{s_2^2}{n_2}}$, is .48. (You do not need to check this calculation.)
  (a) Use these data to construct a 95% confidence interval for $\mu_1 - \mu_2$.
  (b) Based on your CI from part (a), is there evidence, at the .05 level of significance, that the population averages differ? Why or why not?

Answer
  (a) The confidence interval is given by $(6.90 - 6.08) \pm 2.048 \times .48$ or $(-.16, 1.80)$.
  (b) The confidence interval includes zero, so we do not reject H_0 at the .05 level.

32. (Based on an article in *The Plain Dealer* on 12/17/95.) In a sample of 500 American women, 52% said that they would vote Democratic if elections for Congress were held today. This compares to 39% of a sample of 500 men who said they would vote Democratic. Use these data to construct a 90% confidence interval for the difference in population percentages between women and men.

Answer

The standard error is given by $\sqrt{\dfrac{.52 \times .48}{500} + \dfrac{.39 \times .61}{500}}$ = .031. The confidence interval is given by $(.52 - .39) \pm 1.645 \times .031$ or $(.078, .181)$.

33. (Based on a project by Mike Elliot (Oberlin College '97).) Suppose we check to see how many students at a certain college are logged into the computer system at 3 p.m. on each of 75 days and that the sample average is 79, with a sample SD of 13. Further, suppose the data follow a normal curve. The standard error for the sample average is thus $13/\sqrt{75} = 1.5$. For each of the following, state whether it is true or false and explain why. (*Note:* I rounded off a bit in my calculations—do not worry about using 1.96 versus using 2 in a calculation.)

(a) We can estimate that on roughly 95% of all days there will be between 76 and 82 students logged in at 3 p.m.

(b) A 95% confidence interval for the number of students logged in at 3 p.m. on the average day goes from 76 to 82.

(c) If someone takes a random sample of 75 days and goes 2 *SE*s either way from the sample average, there is a 95% chance that this interval will cover the average number of students logged in at 3 p.m. on all days.

(d) If someone takes a random sample of 75 days and uses the sample data to compute a 95% confidence interval for the average number of students logged in at 3 p.m. on all days, there is a 95% chance that this interval will contain 79.

Answer

(a) False. Roughly 95% of the data are in the range given by the mean plus or minus twice the standard deviation, not the mean plus or minus twice the standard error.

(b) True. The confidence interval is given by the mean plus or minus twice the standard error.

(c) True. The range given by the population mean plus or minus twice the standard error covers the sample mean 95% of the time, so the sample mean is within $2 \times SE$ of the population mean 95% of the time.

(d) False. This would only be true if the population mean were 79. However, we do not know what the population mean is.

34. (Based on an article in the 2/11/96 Cleveland *Plain Dealer*.) A group of 1003 Ohio registered voters were asked "How important is it that the federal government has a balanced budget?" From the choices given, 702 of the 1003 chose "very important."

(a) Use these data to construct an appropriate 95% confidence interval.
(b) What assumptions are necessary for the CI in part (a) to be valid?
(c) Interpret, in the context of the setting, your CI from part (a). (What, specifically, is it a confidence interval for? What do the numbers from the CI mean?)

Answer
(a) The sample percentage is $\frac{702}{1003} = .70$. The standard error

is given by $\sqrt{\frac{.7 \times .3}{1003}} = .014$. The CI is given by $(.70) \pm$

$1.96 \times .014$ or $(.672, .728)$.
(b) The data need to be a random sample from the population.
(c) We are 95% confident that between 67.2% and 72.8% of all registered voters in Ohio think that it is very important that the federal government have a balanced budget.

35. The other day I did a two-sample t-test of the hypotheses $H_0: \mu_1 = \mu_2$ versus $H_A: \mu_1 \neq \mu_2$, using samples sizes of $n_1 = n_2 = 12$. The p-value for the test was .12 and α was .10. Suppose that everything except the sample sizes were the same (i.e., $\bar{y}_1$ was the same, $\bar{y}_2$ was the same, etc.), but the sample sizes had been $n_1 = n_2 = 7$. Then:
(i)   the p-value would still be .12 and α would be less than .10.
(ii)   the p-value would still be .12 and α would be greater than .10.
(iii) the p-value would be less than .12 and α would still be .10.
(iv)   the p-value would be less than .12 and α would be less than .10.
(v)   the p-value would be less than .12 and α would be greater than .10.
(vi)   the p-value would be greater than .12 and α would still be .10.
(vii) the p-value would be greater than .12 and α would be less than .10.
(viii) the p-value would be greater than .12 and α would be greater than .10.

Choose one of these and explain why your choice is true.

Answer

(vi)   The choice of alpha level to use does not depend on the sample size. However, as the sample sizes go down, the standard error for the difference in sample means goes up. This means that the *t*-statistic will become smaller (closer to zero), which means that the *p*-value will go up. Also, with smaller sample sizes we have fewer degrees of freedom, which makes the *p*-value larger. Thus, the correct answer is (vi).

36. The other day I did a two-sample *t*-test of the hypotheses $H_0: \mu_1 = \mu_2$ versus $H_A: \mu_1 \neq \mu_2$, using sample sizes of $n_1 = n_2 = 12$. The *p*-value for the test was was .12 and α was .10. It happened that $\bar{y}_1$ was greater than $\bar{y}_2$. Unbeknownst to me, my friend Susan was interested in the same data. However, Susan had reason to believe, based on an earlier study of which I was not aware, that either $\mu_1 = \mu_2$ or else $\mu_1 > \mu_2$. Thus, Susan did a test of the hypotheses $H_0: \mu_1 = \mu_2$ versus $H_A: \mu_1 > \mu_2$. So, for Susan's test:

(i)   the *p*-value would still be .12 and H_0 would not be rejected if $\alpha = .10$.

(ii)   the *p*-value would still be .12 and H_0 would be rejected if $\alpha = .10$.

(iii)   the *p*-value would be less than .12 and H_0 would not be rejected if $\alpha = .10$.

(iv)   the *p*-value would be less than .12 and H_0 would be rejected if $\alpha = .10$.

(v)   the *p*-value would be greater than .12 and H_0 would not be rejected if $\alpha = .10$.

(vi)   the *p*-value would be greater than .12 and H_0 would be rejected if $\alpha = .10$.

Choose one of these and explain why your choice is true.

Answer

(iv)   The choice of alpha does not depend on H_A. For a one-sided test, with the data pointing in the direction predicted by H_A, the *p*-value from the two-sided test is cut in half. Thus, the *p*-value is .06 and H_0 would be rejected.

37. A random sample of 50 subjects received biofeedback training to reduce blood pressure. Researchers measured the decrease in blood pressure for each of them; the average decrease was 11.4 and SD was 1.3. A second sample of 40 control subjects had an average decrease in blood pressure during the study of 5.0, with an SD of 1.4.

(a) Construct a 95% confidence interval for the population difference average in blood pressure decrease between treatment (biofeedback) and control patients.

(b) True or False. Consider just the treatment (biofeedback) group. We can estimate that roughly 95% of all persons given biofeedback training will experience a decrease in blood pressure in the range 11.4 ± 2(1.3), that is, between 8.8 and 14.0. Explain your reasoning.

Answer

(a) The standard error is given by $\sqrt{\dfrac{1.32}{50} + \dfrac{1.42}{40}} = .285$. The CI is given by $(11.4 - 5.0) \pm 2.02 \times .285$ or (5.82, 6.98).

(b) True. The range given by the sample mean ± 2 × SD covers roughly 95% of the data.

38. (Based on data collected by Candace Sady, Oberlin College '96.) Randomly chosen Oberlin students were asked if they had ever used the Oberlin security escort service. Eight out of 28 women said yes, compared to 6 out of 32 men.

(a) Use these data to construct a 95% confidence interval for the difference in population proportions.

(b) Use your CI from part (a) to test, using $\alpha = .05$, H_0: $p_{\text{women}} = p_{\text{men}}$ against H_A: $p_{\text{women}} \neq p_{\text{men}}$. Do you reject H_0? Why or why not?

Answer

(a) The sample percentages are 28.6% and 18.8%. Thus, the standard error is given by $\sqrt{\dfrac{.286 \times .714}{28} + \dfrac{.188 \times .812}{32}} =$ .11. The CI is given by $(.286 - .188) \pm 1.96 \times .11$ or (−.118, .314).

(b) The confidence interval includes zero, so we do not reject H_0 at the .05 level.

39. For each of the following two settings, state the type of analysis you would conduct (e.g., one-sample *t*-test, regression, chi-square goodness-of-fit test, etc.)—be specific—and specify the roles of the variable(s) on which you would perform the analysis, but do not actually carry out the analysis. Also, state the assumptions that are necessary in each case.

(a) Steve Van Geem (Oberlin College '98) collected data on the height (in inches) and shoe size of 14 Oberlin women. Here are the data.

Height	Shoe Size
67	9
65.5	8
67	8.5
67	8.5
64	7.5
65	7.5
67	7.5
64	8
60	6.5
65	7
66.5	9.5
62.5	7.5
68	9
65	8

(b) Ellen Chung (Oberlin College '97) asked a random sample of college students how much they spent on their telephone bill for the month of October, 1995. Here is a summary of the data.

	Men	Women
n	32	31
Mean	26.6	31.2
SD	35.0	20.3

Answer

(a) We could make a scatterplot of the data and find the correlation coefficient between the two variables. We could also fit a regression model using height, X, to predict shoe size, Y.

(b) We could conduct a two-sample t-test and construct a confidence interval for the difference in the population means for men and women.

40. In the spring of 1995, Emily Norland (Oberlin College '95) measured diameter at breast height (DBH), in cm, for samples of the American Sycamore trees on the Vermilion and Black Rivers. Here are the data.

	Vermilion	Black
n	78	46
Mean	35	48
SD	20	26

(a) Use the data to test the null hypothesis that there is no difference in the two population means. Use a two-sided alternative and use $\alpha = .05$. Provide all steps including the value of the test statistic, degrees of freedom, bounds on the p-value, and your conclusion regarding H_0.

(b) In non-technical language, explain to a non-statistician what this means about American Sycamore trees on the two rivers. Be specific.

Answer

(a) The standard error (using the unpooled method) is given by $\sqrt{\dfrac{20^2}{78} + \dfrac{26^2}{46}} = 4.45$. The t-statistic is $\dfrac{35 - 48}{4.45} =$ -2.92. There are 45 degrees of freedom for the test using the unpooled method. Using the pooled method we get $t = $ -3.12 with 122 degrees of freedom. In either case, the p-value is less than .01, so we reject H_0.

(b) We reject the claim that the average diameter at breast height is the same in the populations of American Sycamore trees along the two rivers.

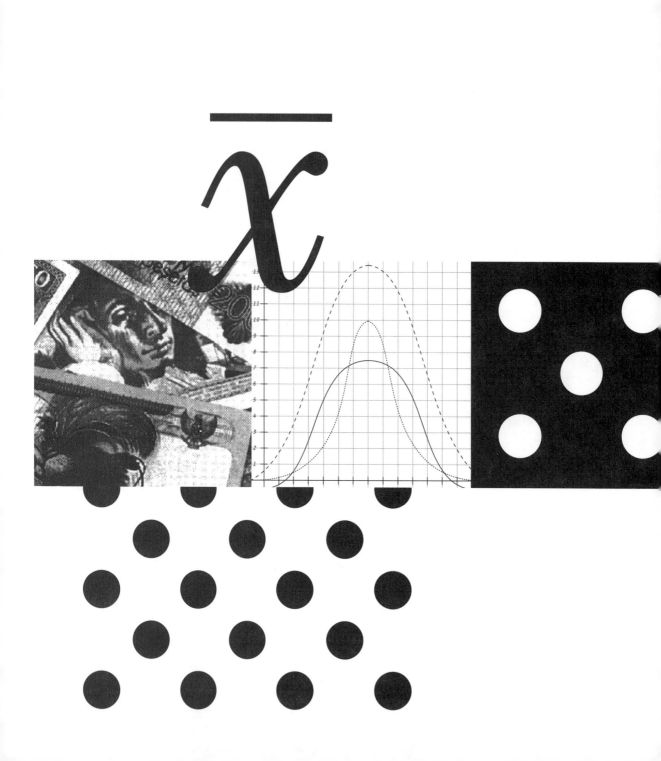

Statistical Process Control

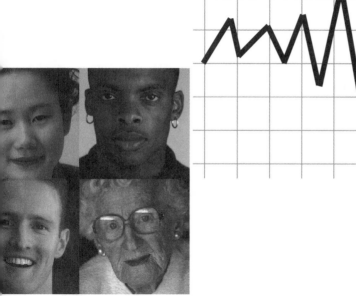

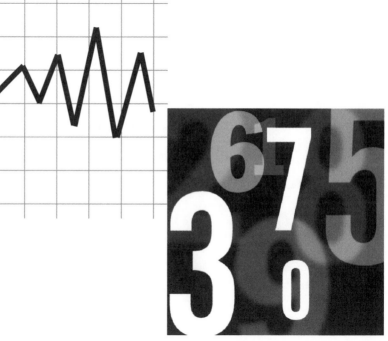

Summary _____

11.1 IN A NUTSHELL

Where does this topic fit in? The ideas of Statistical Process Control (SPC) are not often used in scientific research, nor are they a starting point for other topics in basic statistics. For those reasons SPC is often omitted from a first course. On the other hand, if you are interested in applying statistical ideas to business, SPC is a very important topic. It was invented as a way of improving quality and efficiency in manufacturing, and that remains its main use today. Moreover, many of the ideas of SPC have proved themselves more broadly useful. Things that SPC emphasizes—making systematic use of data to track variability, and shifting attention from discarding defective products to making sure the process is working effectively—are increasingly recognized as important ideas for managing just about any system.

What to learn first For a basic understanding, you should concentrate on three things.

- What is the goal of SPC, and why is it important?
- What is a control chart, and how do you draw one?
- How do you use a control chart to decide whether a process is in control?

Once you've learned these three, it's easy to create variations on the basic control chart using other summary statistics.

What is the goal of SPC, and why is it important? The goal of statistical process control is to detect problems in a manufacturing process *as they arise* so you can make immediate corrections to the process itself. The alternative is to rely on weeding out defective products by inspecting them at the end of the process. This alternative strategy ends up costing a lot more, in part because you don't find out that the process has gone out of control until the final inspection, by which time you've produced many more defective items. The main tool of SPC is the **control chart,** a specialized sequence plot (Topic 4) used to track variation in a manufacturing process, and, in particular, to distinguish two types of variation. The first is chance-like variation, which is always present and is due to **common causes.** The second is the variation due to **special causes,** which makes the process go **out of control.**

What is a control chart, and how do you draw one? To set up a control chart for a process, you follow a standard sequence of steps.

Step 1. *Subgroups.* Decide how you will sample from the process in batches, and how you will group these into subgroups for plotting on the chart. For example, if your process makes potato chips and you want to track salt content, you might take four batches of potato chips from the production line every hour, analyze each batch for salt content, and use the set of four salt values as a subgroup.

Step 2. *Summaries.* Decide what summary statistic you will use. For example, to track average salt level, you would use the mean of the four readings. To track within-batch variability, you could use the range or SD as your summary. If the variable you want to track is categorical, you might use the proportion in a category, such as the fraction of defective items, as your summary.

Step 3. *Center line.* Locate the center line on your graph. For a mean ($\bar{y}$) chart, the center line will be at the average of the subgroup averages.

Step 4. *Control limits.* Compute values of the upper and lower control limits. These values are based on the sampling distribution of the subgroup summaries, and are chosen so that for a process that is in control, the chance of a value outside the control limits is very tiny. (For example, setting the limits for a mean chart at the overall mean $\pm$ 3 SEs will make the chance of a false alarm about 3 in 1000.) That way, you can reasonably conclude that a value outside the control limits is telling you that your process has gone out of control.

Step 5. *Plot.* Once you've marked the center line and control limits on the vertical axis, you just plot subgroup summaries in sequence as you get them, and look for out-of-control signals.

How do you use a control chart to decide whether a process is in control? A process can go out of control in a variety of ways. You could have a single, isolated point outside the control limits, due, perhaps, to a bad batch of raw materials. A change to a new supplier might result in a level of shift, with a new mean for the process. If a machine gradually drifts out of adjustment, you might see a time trend in the means, or an increase in the variability. The simplest out-of-control signals are provided by the upper and lower control limits: a point outside these limits signals that the process is out of control. Other signals are also used.

For example, 2 out of any 3 consecutive points more than 2 SEs away from the center line on the same side, or 9 points in a row (or 8 out of 9) on the same side of the center line.

11.2 VARIATIONS ON THE BASIC IDEA

The mean ($\bar{y}$) chart The subgroup summary is the subgroup mean. The center line (CL) is at the mean of these subgroup means. The control limits are at CL $\pm$ 3 $\times$ SE, where *SE* is the estimated standard deviation of the subgroup mean. (Modern computer programs such as Minitab first compute a pooled SD from the individual observations, then multiply by an unbiasing factor.)

The range (r) chart The subgroup summary is the range of the values in the subgroup. The center line (CL) is at the mean of these subgroup ranges, or, in some computer programs, at a value computed from the pooled standard deviation of the individual observations. The control limits are at CL $\pm$ 3 $\times$ SE, where *SE* is the estimated standard deviation of the subgroup range, taken from a table. (There is no simple formula.)

The standard deviation (s) chart The subgroup summary is the standard deviation of the values in the subgroup. The center line (CL) is at the mean of these subgroup SDs, or, in some computer programs, at a value computed from the pooled standard deviation of the individual observations. The control limits are at CL $\pm$ 3 $\times$ SE, where *SE* is the estimated standard deviation of the subgroup standard deviation, computed as a known multiple of the pooled standard deviation.

The sample proportion (p) chart The subgroup summary is the fraction of defective items in the subgroup. The center line (CL) is at the mean of these subgroup values. The control limits are at CL $\pm$ 3 $\times$ SE, where *SE* is the estimated standard deviation of the subgroup proportion, computed (in the usual way) as $SE = \sqrt{\dfrac{\hat{p}(1 - \hat{p})}{n}}$, where n is the subgroup size and $\hat{p}$ is the proportion of defectives in the set of all subgroups.

Self-Testing Questions

1. Here are data on 18 pieces that come from a sample from an assembly line.

Piece	1	2	3	4	5	6	7	8	9	10	11	12	13	14	15	16	17	18
Number of defects	6	14	7	13	2	3	9	5	0	15	19	6	14	19	13	21	22	14

The first 9 pieces are used to establish the control limits. The SD of the first 9 observations is 4.5 and the mean is 6.6. What are the upper and lower control limits?

Answer
UCL = $6.6 + 3 \times 4.5 = 20.1$;
LCL = $6.6 - 3 \times 4.5 = -6.9$

2. Make a control chart for the data in Question 1.

Answer

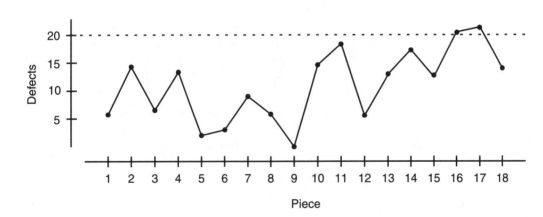

3. Consider the chart from Question 2. Does the chart indicate that the process goes out of control? If so, at what point does this happen?

➤ **Solution**

The process goes out of control at point 16. Prior to that we never have a point outside the control limits, nor do we have 2 out of 3 points in a row higher than the center line plus 2 × SD.

4. A sample of 10 subgroups each of size 50 were selected from a process. Each item was classified as being either defective or not. There were 45 total defectives in the collection, for a proportion of $\dfrac{45}{10 \times 50} = .09$. What are the upper and lower control limits for a p-chart?

Answer

$$\text{UCL} = .09 + 3 \times \sqrt{\frac{.09 \times .91}{50}} = .09 + 3 \times .04 = .09 + .12 = .21;$$

$$\text{LCL} = .09 - .12 = -.03 \text{ (so the effective LCL is 0)}$$

5. Here are proportions of defectives in 9 subgroups of size 50 from the situation described in Question 4.

Subgroup number	1	2	3	4	5	6	7	8	9
% defective	.06	.14	.18	.08	.16	.06	.12	.04	.06

Make a p-chart of these data.

Answer

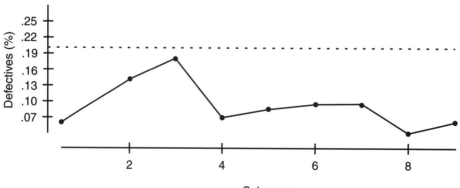

6. Consider the chart from Question 5. Does the chart indicate that the process goes out of control? If so, at what point does this happen?

Answer

The process is in control for the entire chart.

7. An organization has been using control charts on the weight of a certain product, and it has established control limits based on a stable process. Now suppose that a change in supplier of raw materials has the effect of increasing the mean level, but not the variability, of the weight of the product. Will the mean chart detect the change? Will the range chart detect the change?

Answer

The mean chart will detect the change, but the range chart will not be affected.

8. A manager monitors a manufacturing process by using a *p*-chart to record whether or not a batch of products contains defective items. He examines consecutive batches of size 40 and counts the number of defectives in each batch. After 20 batches are examined, he has found a total of 24 defective items. If the process is in control, what would be the best guess at the long-run process defect rate?

Answer

$$.03 = \frac{24}{40 \times 20}$$

9. A baseball team has the following data for the first 8 games of their season.

Game	Hits	At-bats	Proportion
1	3	30	.100
2	14	41	.341
3	5	35	.143
4	6	33	.182
5	10	36	.278
6	12	39	.308
7	15	41	.366
8	10	37	.270

If we want to make a control chart, we should make a mean control chart of the proportion data, since the proportions are based on differing sample sizes. What are the upper and lower control limits for a mean chart?

Answer

$$UCL = .249 + 3 \times .090 = .519$$
$$LCL = .249 - .270 = -.021 \text{ (so the effective LCL is 0)}$$

10. Here are data for games 9–16 for the team from Question 9.

Game	Hits	At-bats	Proportion
9	13	49	.315
10	5	32	.156
11	10	37	.270
12	6	33	.182
13	12	38	.316
14	13	40	.325
15	15	41	.366
16	17	44	.386

Make a control chart of these data.

Answer

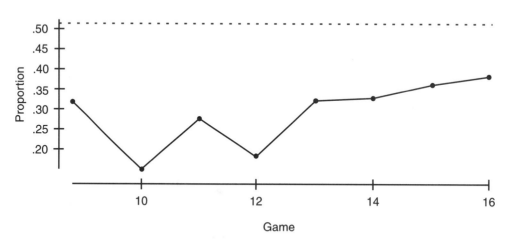

11. Consider the chart from Question 10. Does the chart indicate that the process goes out of control? If so, at what point does this happen?

Answer

The process is in control for the entire chart.

12. Consider the team from Question 9. In a typical 9-inning game, the number of at-bats equals the number of hits + 27. How many hits would the team need to get in order to exceed the upper control limit?

➤ ***Solution***

$30 \times \dfrac{x}{x + 27} > .519$ means that $x > 29.1$, so $x = 30$.

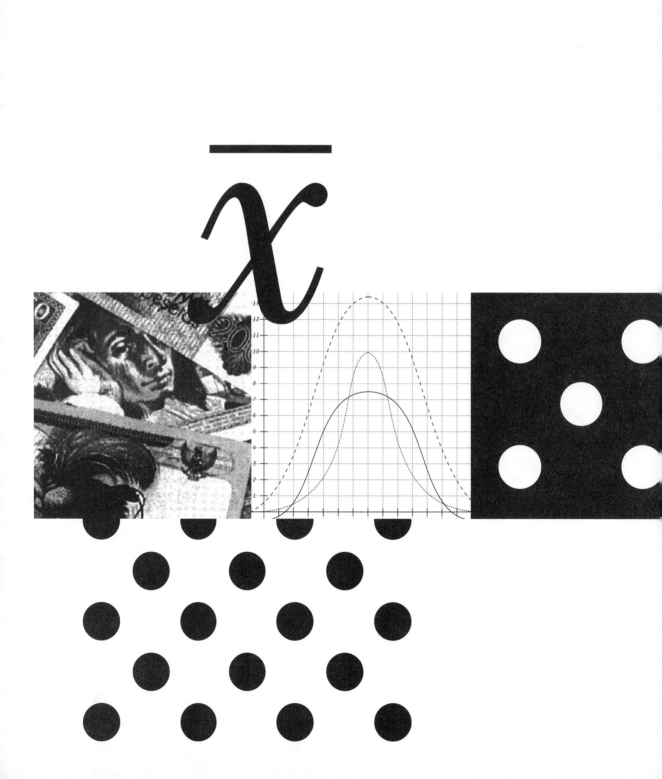

Inference for Relationships

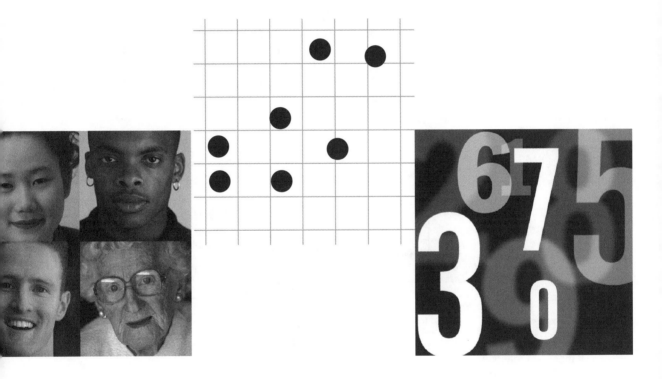

Summary

INTRODUCTION

Most textbooks take two or more chapters for the material presented here. Don't feel you have to cover the whole topic at once. This review is designed so you can do just regression, or just analysis of variance, or just two-way tables of counts (contingency tables), without doing the others first. Feel free to skip around.

This workbook follows David Moore (*Introduction to the Practice of Statistics, Basic Practice of Statistics*) and others who organize the topics of a first course to fit the modern distinction between exploration and inference. This means that if your textbook follows a more traditional organization (e.g., Mario Triola's *Elementary Statistics* or Robert Johnson's *Elementary Statistics*), you'll need to work through the related sections of Topic 5 in this workbook before you start this topic.

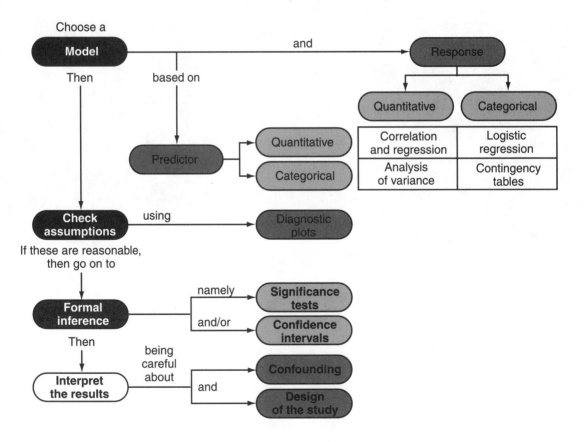

All the methods of inference presented here follow the same general pattern (shown in the diagram on page 276), although details of the assumption-checking, tests, and confidence intervals depend on whether your variables are categorical or quantitative.

12.1 INFERENCE FOR LINEAR REGRESSION

Response and Predictor Are Both Quantitative

Where does this topic fit in? If you know about scatter plots, correlation, and the least squares line (from Topic 5) and about tests and confidence intervals (Topics 9 and 10), then the main things that are new are:

1. the assumptions of the regression model and the diagnostic plots used to check those assumptions; and
2. constructing and interpreting the tests and interval estimates in the regression setting.

What are the main assumptions required for inference, how can you check them, and what can you do if they don't fit your data? Here are five things you should check before you trust the results of your tests and confidence intervals.

1. *Linearity:* If a scatter plot of your data suggests a curve rather than a line or a balloon, then the methods of inference, which are based on balloon summaries, are not suitable. Clear departures from linearity will show up in the plot of y versus x, but as a rule, departures from linearity show up more clearly in a plot of residuals versus fitted values. Often changing x or y or both to a new scale can straighten a curved plot. (An alternative is to use methods for fitting curves.)
2. *Same SDs:* If the SD for the response is not constant and depends on the predictor x, then the formula for the SE of the slope will be wrong, and so will inferences that use that SE. If your residual plot suggests a fan lying along the horizontal axis, with points toward the right more spread out in the vertical direction than points toward the left (or the reverse pattern, with points more spread at the left), then the SD (spread), instead of being constant, depends on the size of the fitted value. For such data sets, a change of scale can often make the SDs in the new scale more nearly constant.
3. *Normality:* Outliers or strong skewness distort the sampling distribution in the sense that the actual sampling distribution is not the one that the methods of inference rely on. Outliers and extreme skewness can be detected in dot plots, box plots,

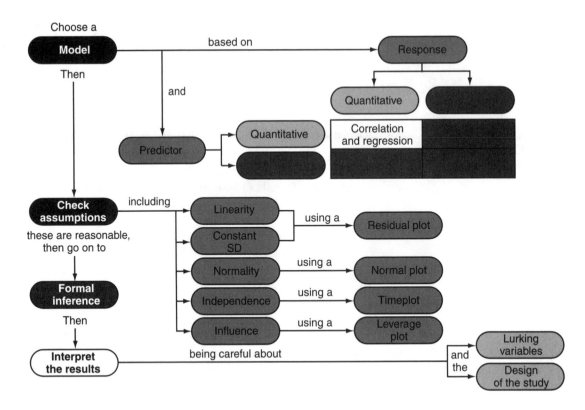

or histograms of your residuals. For a more sensitive check of the normality assumption, use a **normal quantile plot**. Here, as with the first two assumptions, a change of scale is often the best remedy: Changing the scale of the response can make a skewed distribution more nearly symmetric. (See Topic 2.)

4. *Independence:* If your observations are correlated with each other, the estimated SD will be off and so will any *SE*s computed from the SD. There are a variety of ways to check this assumption, depending on your particular situation. One simple and useful check, for observations taken in sequence, is a time plot (Topic 4). Patterns in the time plot suggest dependencies, which can invalidate inferences based on the methods of this chapter. In particular, time series data often shows such dependencies. For such data, you should rely on the (more advanced) methods of inference for time series.

5. *Leverage:* The farther a predictor value *x* is from the mean of the *x*-values, the more influence the corresponding point has on the fitted slope. A scatterplot can show extreme instances of high leverage points; a more sensitive check is to use a computer for a sequence plot of leverage values (labeled HI in

Minitab). Ordinarily, you shouldn't trust a fitted line that is heavily influenced by a single isolated point. Sometimes transforming the predictor to a new scale will help. If a point still has high leverage even after a change of scale, you can refit the line without the influential point to see how much effect it has.

The details of the methods If you have a good grasp of the ideas of tests and confidence intervals already, the methods here should seem quite familiar in their general pattern. **Inference for the slope** of a least squares line follows the same pattern as inference for means and proportions (Topic 10). A **confidence interval** has the form $\hat{\beta} \pm t^* \times SE(\hat{\beta})$, where t^* comes from a t-table with degrees of freedom $n - 2$.

For a **significance test** of the null hypothesis that the true slope equals β_0, compare $\dfrac{\hat{\beta} - \beta_0}{SE(\hat{\beta})}$ with t^*.

To test that the true slope $= 0$, use $\hat{\beta} / SE(\hat{\beta})$. For **predicted values** there are two confidence intervals, depending on whether your goal is to predict the mean response $\hat{\mu}$ for a fixed value of the predictor, or to predict an individual observation. The predicted value for the mean is less variable than the predicted value for an individual observation.

How do you interpret the results of your inferences? With caution! Interpretation can be tricky, especially when your data, instead of being experimental, as in the physical sciences, are observational, as is so often true in economics or sociology. Be sure to plot your data, and keep in mind the Anscombe plots (on the CD), which show four very different patterns for data sets with identical summary statistics. All the cautions from Topic 5 about jumping to conclusions based on high correlations still apply.

- *Extrapolation:* Don't assume that an observed relationship holds outside the range covered by your data.
- *Lurking variables:* The x,y relationship you see may involve some unobserved third variable.
- *Ecological fallacy:* A relationship that holds for group averages may not hold for the individuals in the groups.

Above all, remember that *association is not causation*.

12.2 INFERENCE FOR COMPARING GROUPS, OR ANALYSIS OF VARIANCE (ANOVA)

Response Is Quantitative, Predictor Is Categorical

Where does this topic fit in? If you know about parallel plots for comparing groups (Topic 5), and about percentage of variation accounted for (Topic 5), and about tests and confidence intervals (Topics 9 and 10), then the main things that are new are

1. the assumptions of the ANOVA model and the diagnostic plots you can use to check them;
2. decomposing a data set into "overlays" and comparing mean squares to test hypotheses;
3. linear contrasts; and
4. adjustments for multiple comparisons.

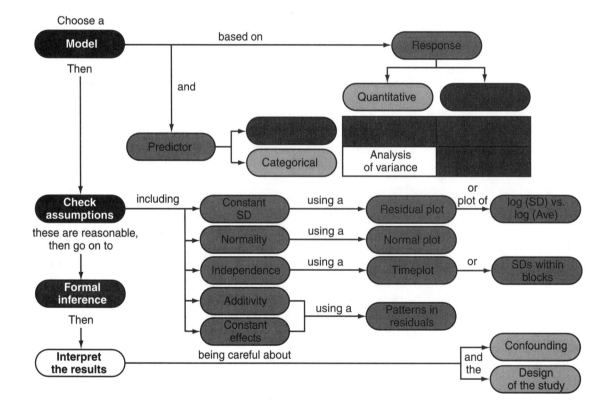

What are the main assumptions required for inference, how can you check them, and what can you do if they don't fit your data? Here are four things you should check before you trust the results of your tests and confidence intervals.

1. *Constant, additive effects:* ANOVA is based on comparing averages. For this approach to make sense, the averages must correspond to meaningful features of the data. For example, it only makes sense to estimate "the" effect of a treatment if there's one constant effect for that treatment. If the effect of the treatment varies a lot from one experimental unit to the next, or if the effects are not additive (they might instead exert a constant *percentage* effect, for example), then averages won't give you meaningful comparisons. Often the best check of these assumptions is indirect: if they fit your data, then the other assumptions will also, and a check of those assumptions will tell you that it's OK to rely on the methods of inference.

2. *Same SDs:* If the SD for the response is not constant (for example, if there is a different SD for each group), then the actual significance levels and confidence levels will be different from the ones the methods say you are getting. You can check this assumption from a plot of residuals versus fitted values (which for ANOVA looks like a parallel dot graph): are the spreads of the groups roughly equal? You can also compute and compare group SDs. If spreads are unequal, with the size of the spread related to the size of the average, transforming to a new scale will often make the spreads more nearly equal.

3. *Normality:* Outliers or strong skewness distort the sampling distribution in the sense that the actual sampling distribution is not the one that the methods of inference rely on. Outliers and extreme skewness can be detected in parallel dot plots or box plots, or sometimes from a histogram of your residuals. For a more sensitive check of the normality assumption, use a **normal quantile plot**. A plot of normal-shaped residuals will suggest a line; skewness and long tails give curved plots; outliers appear as isolated points off the line. Here, as with the assumption of equal SDs, a change of scale is often the best remedy.

4. *Independence:* If your observations are correlated with each other, the estimated SD will be off and so will any *SE*s computed from the SD. There are a variety of ways to check this assumption, depending on your particular situation. One simple and useful check, for observations taken in sequence, is a time plot (Topic 4). Patterns in the time plot suggest dependencies, which can invalidate inferences based on the methods of this chapter.

How to decompose a data set Analysis of variance is based on the assumption that each data value behaves as though it had been built as a sum of pieces: one piece for the grand mean (common to all observations), an additional piece for the effect of the group the data value belongs to (the treatment group), plus a chance error due to the random assignment of treatments to experimental units. This sum of terms, together with the assumptions listed above, is the **model** for one-way ANOVA. Analyzing the data starts by splitting each data value into a sum of pieces to try to recapture (estimate) the pieces in the model. Thus each data value gets split into the overall grand average of all the observations (an estimate of the grand mean or common term), an estimated group effect (equal to the group average minus the grand average), and a residual. The sum of the first two pieces (grand average plus estimated treatment effect) is the fitted value; the residual, as always, equals observed minus fitted. This splitting into pieces lets us write the whole data set as a sum of three overlays, one for each term in the model, as shown in the CD.

Degrees of freedom, mean squares, and *F*-ratios Just about any textbook that covers analysis of variance will tell you how to compute **degrees of freedom (df)**, and how to use df in constructing your test statistics. The concept, unfortunately, is hard to understand. This short summary concentrates on an informal explanation of why you have to bother about degrees of freedom. When you decompose a data set, you end up with more estimated terms than observed values: you have a grand mean, one group mean for each group, and one residual for each data value. The information in the original data set has been redistributed, but you haven't created any new information. Rather, there are redundancies in your estimates because all the estimates but the grand mean are deviations from one average or another, and the deviations from an average always add to 0. When you compute a test statistic in analysis of variance, you form an **F-ratio**—a ratio of two measures of average variability—called **mean squares**. For one-way ANOVA, the numerator mean square measures the average variability due to treatments (i.e., from one group mean to another); the denominator mean square measures the average variability in the residuals. If the ratio is near 1, we conclude that the variability in the treatment means is no bigger than the variability in the residuals: the treatment differences could be due just to chance. On the other hand, if the F-ratio is large enough, we conclude that the estimated effect of the treatments is too big to be due just to chance. In order for this logic to work, our measures of average variability (sum of squares divided by degrees of freedom) need to take into account the redundancies in the estimates that you square and add to get the sums of squares. Dividing by df instead of the number of estimates makes just the right adjustment.

Estimated Effect	Degrees of Freedom
Grand mean	1
Group mean – grand mean	# groups – 1
Residual = data value – group mean	# data values – # groups

Comparisons and contrasts In ANOVA, you use the F-ratio to test whether there are any differences anywhere among the group means. Sometimes you want more focus. Contrasts or comparisons allow you to test hypotheses about particular pairs of means or about other meaningful groups of observations. There are three common strategies for choosing contrasts. **Planned comparisons** arise because you have a natural structure to your groups that tells you which comparisons make sense, or you have some other reason based on the applied context for wanting to compare particular groups. (The CD gives an example.) Instead, you might simply test **all possible pairs**. At other times, the results of the experiment may suggest particular comparisons; these are called **post hoc** (after the fact) and require special methods of adjustment. **Inference for contrasts** follows much the same pattern as for the difference between two means (Topic 10): the test statistic equals the distance from your estimate to the null value, measured in standard errors, and the confidence interval has the form Estimate $\pm t \times SE$. Here are the parts that are new.

- *Estimate:* Write your contrast as a weighted sum of group means: $\text{Est} = \sum c_i \bar{y}_i$.

- *Null hypothesis:* Weighted sum of true means is 0: H_0: $\sum c_i \mu_i = 0$.

- *Standard error:* $SE = s\sqrt{\sum \dfrac{c_i^2}{n_i}}$, where $n_i = $ # data values in group i.

- *Degrees of freedom:* Use $\text{df}_{\text{Res}} = $ # data values $-$ # groups.

Adjusting for multiple comparisons Contrasts typically come in families, as, for example, the set of all possible pairs of groups. Each comparison has a fixed chance α of giving a false alarm, so if you do a lot of them at once, your chance of at least one false alarm (called the "family-wise" error rate) can be quite high. There are several methods for adjusting your comparisons in order to keep the familywise error rate at a fixed level such as 5%. The method you use depends on how you choose your contrasts.

12.3 CONTINGENCY TABLES

Response and Predictor Are Both Categorical

Where does this topic fit in? When both the predictor and the response are categorical, and so your data form a two-way table of counts (a contingency table), the most usual null hypothesis to consider is that there is no association between rows and columns. If you've already read the section in Topic 5 on patterns in two-way tables of counts, then you already know about the model of no association, and the only thing new here is the chi-square test of this null hypothesis.

Assumptions The chi-square test is valid for a number of different probability models. The essential conditions are met if data come from a population or process and each row is based on an SRS, the whole table is based on a single SRS, or the data are for an entire population.

Test statistic To test the hypothesis of no association, you first compute an expected value for each row/column pair, as described in Topic 5: Expected count = (Row total) × (Column percentage). The test statistic is the sum over all the cells of ` (Observed − Expected)^2 / Expected. You compare the value of this test statistic with a chi-square distribution having df = (# rows − 1)(# columns − 1). In order to use the chi-square test, all expected counts should be greater than 1, and no more than 20% of the expected counts should be less than 5.

12.4 LOGISTIC REGRESSION

Response Is Categorical, Predictor Is Quantitative

Where does this topic fit in? If you haven't yet read the (very short!) section (5.4) on logistic regression in Topic 5, now would be a good time to read it. That section explains why logistic regression is important, and why it's usually left out of a first statistics course, even though it deals with one of the four basic families of two-variable relationships.

12.5 MULTIPLE REGRESSION

Response Is Quantitative, There Are Two or More Predictors, and at least One Is Quantitative

Where does this topic fit in? Multiple regression extends simple (one-predictor) linear regression to data sets with more than one predictor variable by expressing the response value as a weighted sum of the values of the predictors. The weights (regression coefficients) are chosen following the same least squares principle from before. Inference also follows the same pattern as for one-variable regression. You can test hypotheses about the values of the fitted regression coefficients or construct confidence intervals, either for the regression coefficients or for predicted values of the response. There are many reasons for using multiple regression: to study the effects of two or more predictors at once, to fit curves (using x and x^2 as predictors) or parallel lines (for example, separate lines for men and women), to adjust for some unwanted influence in order to see an effect more clearly, or simply to get better predictions by reducing the size of the residuals.

An example The CD works through an example in some detail, using the data introduced in the problems of Topic 2, and asking whether academic salaries discriminate against women. The cases are various academic fields (subjects), the response is the average salary, and the predictors include the percent women in the field and the percent unemployment in the field. There is a strong negative correlation between salary and percent women: fields with a high percentage of women have lower average salaries. However, classical economic theory (the law of supply and demand) predicts that in fields where unemployment is higher, salaries will be lower, and vice versa, so it may be that the salary differences are accounted for by a correlation between unemployment rate and salary. Both these explanations (discrimination against women, the law of supply and demand) are based on one-predictor models. Multiple regression makes is possible to study the effect of both predictors at once.

Caution Interpreting the results of a multiple regression analysis requires even more caution than with just one predictor. All the previous cautions of Topics 5 and 12 apply; and, since predictors are often correlated with each other, there are new concerns. The meaning of fitted slopes is based on changing just one predictor at a time, but, as illustrated in the CD, if predictors are correlated, they tend to change together, not one at a time. Moreover, if two predictors are strongly correlated, a scatter plot of the two predictors will be long and skinny. The data will contain little information about the effect on the response of changes to the predictors in the skinny direction, and the corresponding part of the fitted relationship will be unstable.

12.6 TWO-FACTOR ANALYSIS OF VARIANCE: THE COMPLETE BLOCK AND TWO-WAY DESIGNS

Response Is Quantitative, the Two Predictors Are Categorical

Where does this topic fit in? This topic generalizes one-way analysis of variance to data from randomized complete block and two-way factorial designs (Topic 6). For the two-way factorial design, you have two factors of interest and more than one observation per cell in order to measure interaction. For the complete block design, you have two factors but only one is of interest; the other is a nuisance factor. Moreover, this design uses only one observation per cell because the block-by-treatment interaction is assumed to be 0. (*Note:* Be sure you've done the sections on the complete block and two-way designs in Topic 6 and the section on one-way ANOVA in this topic before you do this section.)

The complete block design The model for this design writes each data value as a sum of four pieces: grand mean, block effect, treatment effect, and chance error. An analysis based on this model splits each data value into four corresponding pieces.

Estimated Effect	Degrees of Freedom
1. Grand mean	1
2. Block mean – grand mean	# treatments – 1
3. Treatment mean – grand mean	# treatments – 1
Residual = data value – sum of estimated effects (1)–(3)	# data values – sum of df for effects (1)–(3)

The two-way factorial design The model for this design writes each data value as a sum of five pieces: grand mean, effect for factor 1, effect for factor 2, interaction, and chance error. An analysis based on this model splits each data value into five corresponding pieces.

Estimated Effect	Degrees of Freedom
1. Grand mean	1
2. Row mean – grand mean	# levels for the row factor – 1
3. Column mean – grand mean	# levels for the column factor – 1
4. Cell mean – {grand mean + row effect + col. effect}	# cells – { 1+ df_{Rows}+ df_{Cols} } = $df_{Rows} \times df_{Cols}$
Residual = data value – sum of estimated effects (1)–(4)	# data values – sum of df for effects (1)–(4)

Self-Testing Questions

SIMPLE LINEAR REGRESSION AND CORRELATION

Country	1991	1992
Canada	517	1025
China	1582	897
Czech.	243	144
France	1687	1435
Poland	116	21
Germany	2531	1128

1. The table at left shows the amount of arms exported, in millions of constant (1991) dollars, for several countries for 1991 and 1992.

The fitted regression model is $\hat{y} = 277 + .448x$, where $y = $ 1992 amount and $x = $ 1991 amount. The standard error of the slope is .189. Construct a 95% confidence interval for the population slope.

Answer $.448 \pm 2.776 \times .189$ or $(-.077, .973)$

2. Consider the data from Question 1. Is the slope of the regression line significantly different from 0 at the .05 level of significance?

➤ **Solution**

No, the t-ratio is $\dfrac{.448}{.189} = 2.36$, so the p-value is greater than .05.

3. Here is a residual plot for the regression model from Question 1. The residual plot is helpful in checking which of the regression assumptions?

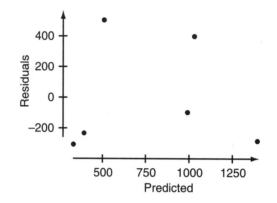

Answer Linearity and Same SDs

4. The scatter plot on p. 289 shows the relationship between the number of post offices in the U.S. (in thousands) and the number of pieces of mail handled (in billions).

Year	# Offices	# Pieces
1988	29.2	161.0
1989	29.1	161.6
1990	29.0	166.3
1991	28.9	165.9
1992	28.8	166.4
1993	28.7	171.2
1994	28.7	177.1

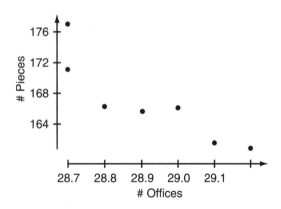

The regression model is $\hat{y} = 895.4 - 25.2x$. The standard error for the slope is 6.1. Find a 95% confidence interval for the slope.

Answer $-25.2 \pm 2.571 \times 6.1$ or $(-40.9, -9.5)$

5. Consider using the data from Question 4 to test the relationship between offices and pieces of mail.
(a) State H_0 in symbols.
(b) Conduct the test. Do you reject H_0 at the .01 level of significance?

➤ ***Solution***
(a) H_0: $\beta = 0$
(b) $t = \dfrac{-25.2}{6.1} = -4.13$. The p-value is less than .01, so we reject H_0.

6. How could we check the assumption of independence for the regression model from Question 4?

➤ ***Solution***
Since the data have a time sequence, we could plot residuals from the regression model against time and look for a pattern.

7. Could we use the model from Question 4 to predict the number of pieces of mail that would be handled if there were only 25,000?

➤ **Solution**

No. 25 is well below the smallest *x*-value in the data set, so it would be dangerous to extrapolate out this far.

8. A politician saw the data and regression model from Question 4 and concluded that closing post offices is a good way to increase the amount of mail handled by the postal service. Comment on this conclusion.

➤ **Solution**

The conclusion does not follow from the data. There is a lurking variable—year—that is probably affecting the increase in the amount of mail being handled. We cannot conclude that there is a cause-and-effect relationship between *x* and *y* in this situation.

9. Fill in the blank. The larger the standard error for the slope, the _____ (wider/narrower) the 90% confidence interval for the slope.

Answer Wider

10. The following scatter plot on p. 291 shows the relationship between two major economic indicators.

Dow Jones Industrial Average	American Stock Exchange Index
76.6	68.1
65.8	74.0
71.5	68.9
77.8	92.6
72.9	92.5
83.7	108.1
95.5	136.0
91.3	161.8
94.2	150.0
93.0	180.1
98.9	183.2
103.9	206.4
109.8	229.0
105.6	243.7

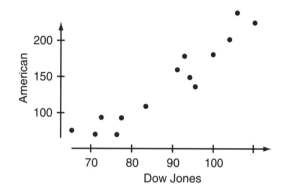

The fitted regression model for these data is $y = -221.1 + 4.1x$. The standard error of the slope is .4. Consider testing $H_0: \beta = 0$. State the null hypothesis in words, in the context of this setting.

Answer

H_0: The Dow Jones Index is of no value in predicting the American Stock Exchange Index.

11. **(a)** What is the value of the test statistic for the hypothesis test from Question 10?
(b) How many degrees of freedom does the test statistic have?

Answer
(a) $t = \dfrac{4.1}{0.4} = 12.25$
(b) The t-statistic has 12 degrees of freedom.

12. How could we check the normality assumption that is needed for the test from Question 11?

Answer

We could make a normal plot of the residuals or look at a histogram of the residuals.

13. The following residual plot from the regression model in Question 10 raises some doubts about which of the regression assumptions?

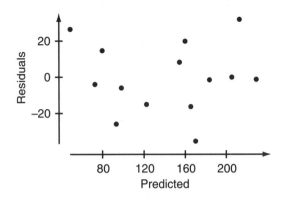

Answer

The residual plot suggests that the relationship is not linear. Rather, there seem to be a bit of a curve in the relationship.

14. The correlation coefficient between the Dow Jones Index and the American Stock Exchange Index for the data in Question 10 is .95. Find and interpret the value of r^2.

➤ **Solution**

$r^2 = .95^2 = .903$. Thus, we can account for 90.3% of the variability in the American Stock Exchange Index by using the Dow Jones Index in a regression model.

ANALYSIS OF VARIANCE: ONE FACTOR

Solitary	Control
7	20
16	15
10	17
14	17
5	16
6	13
11	6
8	22
3	21
20	12

15. The data in the table at the left come from a study in which ten prison inmates were kept in solitary confinement and ten others were used as controls. The response variable is the frequency of alpha waves.

Here is a partially completed ANOVA table for these data.

Source	df	SS	MS	F
Groups	1	174.05		
Error	18			
Total	19	634.95		

Complete the ANOVA table.

Answer

Source	df	SS	MS	F
Groups	1	174.05	174.05	6.797
Error	18	460.90	25.61	
Total	19	634.95		

16. State, in symbols, the null hypothesis that is tested in the ANOVA table in Question 15.

Answer $H_0: \mu_1 = \mu_2$

17. Consider the ANOVA table from Question 15. If $\alpha = .10$, do you reject H_0?

➤ ***Solution***

Yes, since the F-ratio is greater than 4.41

18. Here is a normal plot of the residuals from the ANOVA in Question 15. What does this plot indicate about the ANOVA assumptions?

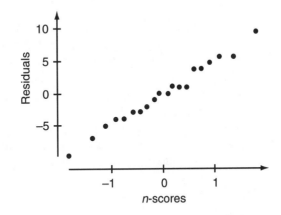

Answer

The plot indicates that the assumption of normality is met. The plot does not address the other assumptions.

19. Here is a residual plot from the ANOVA in Question 15. What does this plot indicate about the ANOVA assumptions?

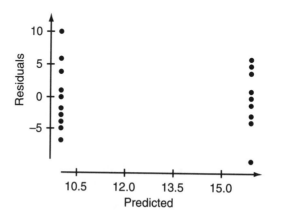

Answer

The plot indicates that the assumption of equal SDs is met. The plot does not address the other assumptions.

20. True or False. On p. 295 there are parallel dot plots for 5 groups from an experiment. The plot indicates that an ANOVA would not be appropriate, since one of the groups clearly has a higher average than the others.

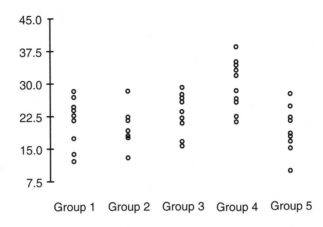

➤ **Solution**

False. ANOVA is designed to test whether, in fact, the difference between group averages might have arisen by chance.

---

21. True or False. Here are parallel dot plots for 5 groups from an experiment. The plot indicates that an ANOVA would be appropriate.

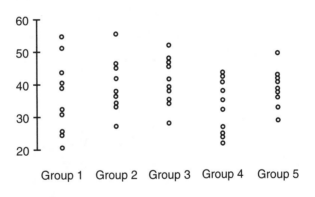

➤ **Solution**

True. The ANOVA assumptions seem to be satisfied here.

---

22. True or False. Here are parallel dot plots for 5 groups from an experiment. The plot indicates that an ANOVA would be appropriate.

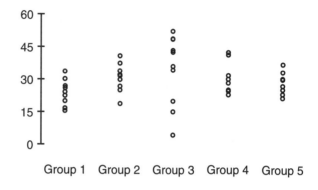

> **Solution**

False. Group 3 seems to have a much higher SD than do the other groups.

---

23. True or False. In doing an ANOVA, if the null hypothesis is true, the numerator and denominator mean squares that go into the F-ratio will be estimates of the same quantity, so the F-ratio will tend to be near 1.

Answer True

---

24. True or False. In doing an ANOVA, there are two possible meanings for a large F-ratio: either the null hypothesis is true or an unlikely event has happened.

> **Solution**

False. A large F-ratio could indicate that the null hypothesis is false, not true.

---

25. Consider an ANOVA that compares 7 groups, with 5 observations taken in each group.
  (a) How many degrees of freedom are there for the groups?
  (b) How many error degrees of freedom are there?

Answer **(a)** 6 **(b)** 28

26. In an ANOVA problem with 3 groups a researcher wanted to compare group 1 to group 3, so she formed the contrast $\bar{y}_1 - \bar{y}_3$. Each sample mean was the average of 10 observations. The values of $\bar{y}_1$ and $\bar{y}_3$ were 12.5 and 8.3, respectively.

(a) What is the value of the contrast?

(b) The value of s (the square root of the MS Error) from the ANOVA was 1.2. What is the standard error of the contrast?

(c) How many degrees of freedom are there for the contrast?

Answer

(a) $12.5 - 8.3 = 4.2$

(b) $SE = 1.2 \times \sqrt{\dfrac{1^2}{10} + \dfrac{(-1)^2}{10}} = 0.537$

(c) $30 - 3 = 27$

27. Researchers conducted an ANOVA to compare several treatment groups. The ANOVA table is given below.

Source	df	SS	MS	F
Groups	3	250.20	83.40	4.57
Error	20	364.75	18.24	
Total	23	614.95		

(a) How many groups were in the experiment?

(b) If $\alpha = .10$, is the null hypothesis rejected?

➤ *Solution*

(a) There were 4 groups, since df for groups is 3.

(b) Yes, we reject H_0 if the .10 cut-off for an F with 3 and 20 degrees of freedom is 2.38.

ANALYSIS OF VARIANCE: MORE THAN ONE FACTOR

28. Consider the following table, which summarizes an analysis of variance of the effect of the month and the phase of the moon on admissions to mental hospitals.

Source	df	SS	MS	F
Months	11	45012	4092	6.85
Phases	2	3744	1872	3.13
Error	22	13140	597.2	
Total	35	61896		

What does this ANOVA tell us about which effect(s) is significant at the .05 level?

➤ **Solution**

The months effect is significant, since 6.85 > 2.26 (the .05 cut-off for an *F* with 11 and 22 degrees of freedom), but the phases effect is not significant, since 3.13 < 3.44 (the .05 cut-off for an *F* with 2 and 22 degrees of freedom).

---

29. The following ANOVA table summarizes the data from an experiment in which the response variable was the weight gain of pigs and the two treatments were (1) whether or not the pig was given a dose of an antibiotic and (2) whether or not the pig was given a dose of Vitamin B12. Complete the ANOVA table.

Source	df	SS	MS	F
Antibiotic	1	192	192	5.30
B12	1	2187	2187	
Interaction	1	1875		51.72
Error	8	290		
Total	11	4544		

Answer

Source	df	SS	MS	F
Antibiotic	1	192	192	5.30
B12	1	2187	2187	60.33
Interaction	1	1875	1875	51.72
Error	8	290	36.25	
Total	11	4544		

30. Consider the ANOVA table from Question 29. The *F*-ratio for the interaction is 51.72, which is a very large value. Thus, the interaction effect is significant at the .01 level. Can we conclude that the antibiotic and Vitamin B12 work well together in increasing a pig's weight?

➤ **Solution**

No. The significant *F*-ratio tells us that the effect that B12 has on the pig depends on whether or not the antibiotic is present. The ANOVA calculations by themselves do not tell us what combination of B12 and antibiotic is best.

31. Here is a residual plot from an analysis of variance in which two factors were studied to see how they affect the rate of leaves falling from a plant. What does this plot tell us about the ANOVA assumptions?

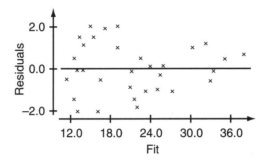

➤ **Solution**

It appears as if the SDs might not be the same for all combinations of the factors since there is less variability in the residuals toward the right side of the plot.

---

32. The following ANOVA table summarizes a study of the brain activity of a set of shrews during different kinds of REM sleep. Each of the 6 shrews in the study acted as a block.

Source	df	SS	MS	F
Shrew (block)		194.16		19.34
Sleep	2	15.24	7.62	3.79
Error	10	20.08	2.01	
Total	17	229.48		

(a) How many degrees of freedom are there for the shrews (blocks)?

(b) Describe what the ANOVA tells us.

➤ **Solution**

(a) 5, since there were 6 shrews.

(b) The effects of the shrews is large, so there is a lot of variability between the shrews, but the effects of the three kinds of sleep is not significant at the .05 level.

---

33. Here are three residual plots from analyses of variance. Do the ANOVA assumptions hold for each of them? If not, what are the problems?

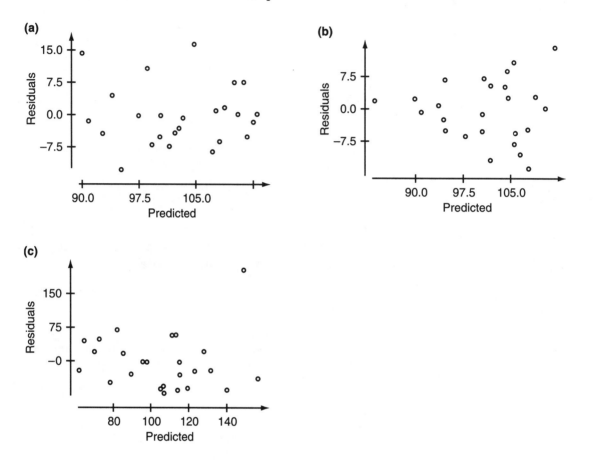

(a)

(b)

(c)

Answer

(a) This plot looks fine.

(b) The wedge shape of the plot suggests that the SDs are not the same for all groups.

(c) There is an outlier here.

34. There are three interaction graphs from a two-way ANOVA shown on page 301. Which of these indicate that there is an interaction present?

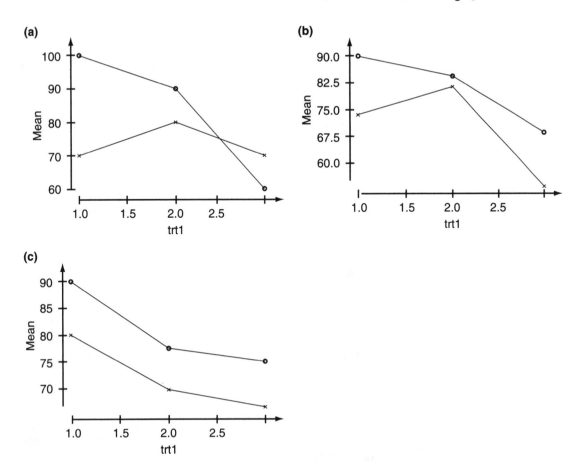

> ➤ *Solution*
>
> **(a)** Yes, there is an interaction.
> **(b)** Yes, there seems to be an interaction, since the lines are not parallel. The lines don't have to cross for there to be an interaction.
> **(c)** No, there does not seem to be an interaction here.

MULTIPLE REGRESSION

35. A group of 50 cars were used to determine a regression model for predicting gallons of gasoline used per 100 miles in terms of horsepower and drive ratio. The equation is

$$GP100M = 4 + .0123 \times \text{horsepower} - .52 \times \text{drive ratio}$$

A new car has not been tested yet. It has a horsepower rating of 130 and a drive ratio of 3.2. What does the regression model predict will be the car's gasoline mileage (in gallons per 100 miles)? (Round to nearest gallon.)

Answer

$4 + .0123 \times 130 - .52 \times 3.2 = 4 + 1.60 + 1.67 = 7.26$, which we would round off to 7.

36. Fill in the blank. The major difference between simple regression models and multiple regression models is that multiple regression models have several _____ (dependent/explanatory) variables.

Answer Explanatory

37. Suppose we give a group of students a test on world cultures and we think that awareness of world cultures is related to how much foreign language instruction the students have had. Consider modeling the test scores, y, by using number of semesters of foreign language study, x, and an indicator variable, z, that is 1 for women and 0 for men. Suppose the multiple regression model is $y = b_0 + b_1 x + b_2 z$. What coefficient or combination of coefficients represents the predicted test score of
(a) a man who has never studied a foreign language?
(b) a women who has studied foreign language for 1 semester?
(c) a women who has never studied a foreign language?

Answer
(a) b_0
(b) $b_0 + b_1 + b_2$
(c) $b_0 + b_2$

38. The following is a graph of height versus hand strength for a sample of college students. The women are graphed as Xs and the men are graphed as Os. What feature of the graph indicates that a multiple regression model, using both height and sex, is appropriate for these data?

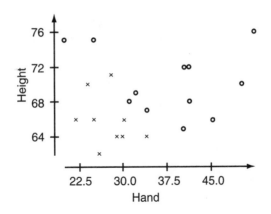

> **Solution**

The men and women are separated on the graph, which suggests that there is a male/female effect.

39. Here is a scatter plot that shows a positive relationship between x and y for juniors, graphed as Os, and for seniors, graphed as Xs. If we ignore year in school and just include all of the data in the simple linear regression model, will the coefficient of x be positive, negative, or 0?

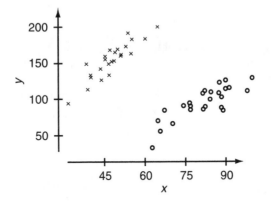

Answer The coefficient will be negative.

40. The regression model whose equation is $y = b_0 + b_1x + b_2z$, in which x is a continuous variable and z is a binary indicator variable, is best described as _____.

> **Solution**

Two parallel lines (one when $z = 1$ and another when $z = 0$).

41. The following are residual plots from two multiple regression models. In each case, what does the plot say about the appropriateness of the assumptions that are used in multiple regression?

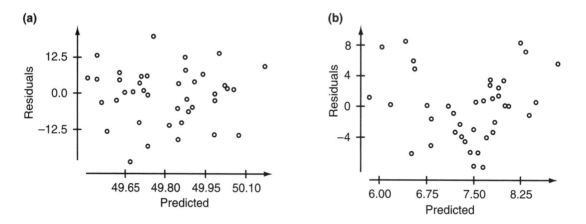

(a)

Residuals

12.5

0.0

−12.5

49.65 49.80 49.95 50.10
Predicted

(b)

Residuals

8

4

0

−4

6.00   6.75   7.50   8.25
Predicted

> **Solution**

(a) Everything looks OK here.

(b) This plot shows that there is curvature in the data, so a quadratic term is called for.

42. Here is a multiple regression model that relates the market values of 77 major companies to their annual sales and their assets: market value = 509.08 + .291 × sales + .044 × assets. The standard error for the assets coefficient is .031. Conduct a test of the null hypothesis that assets are not related to market value, given that we have controlled for sales.

> **Solution**

The *t*-ratio is $\dfrac{.044}{.031}$ = 1.42. With 74 degrees of freedom, this *t*-statistic is not significant, so we do not reject H_0.

CONTINGENCY TABLES

43. The following is a table of rhyme schemes that appeared in the beginning lines of a randomly selected sample of Dr. Suess verses. There were four rhyme schemes and two kinds of books.

	No Rhyme	One Rhymed Pair	Rhymed Triplet	Two Rhymed Pairs	Total
Beginner book	3	25	4	2	34
Non-beginner	7	7	3	18	35
Total	10	32	7	20	69

(a) Based on this sample, what is the estimate of the proportion of the Beginner books that have one rhymed pair?

(b) Based on this sample, what is the estimate of the proportion of the Non-beginner books that have one rhymed pair?

Answer (a) $^{25}/_{34} = .735$ (b) $^{7}/_{35} = .20$

44. Consider the data from Question 43. Assuming independence of rhyme scheme and book type, what is the expected count for Beginner books with a rhymed triplet?

➤ **Solution**

We need to take (row total) × (column total)/(grand total). This is $34 \times {}^{7}/_{69} = 3.45$.

45. Consider the data from Question 43. The chi-square statistic for testing independence is 24.67.

(a) How many degrees of freedom does the test statistic have?

(b) If $\alpha = .05$, do you reject H_0?

➤ **Solution**

(a) There are $3 \times 1 = 3$ degrees of freedom (from (# rows − 1) × (# columns − 1)).

(b) Yes, we reject H_0, since the .05 cut-off for a chi-square with 3 degrees of freedom is 7.81.

46. Mann et al. (*British Medical Journal*, 1975) studied the relationship between smoking and heart attacks. The data they collected are shown in the following table.

	HEART ATTACK?		
Smoker?	**Yes**	**No**	**Total**
Yes	45	83	128
No	14	74	88
Total	59	157	216

What proportion of the smokers had heart attacks?

Answer $45/128 = .352$ or 35.2%

47. Consider the data from Question 46. State, in words, the null hypothesis that is tested with a chi-square test.

Answer

The null hypothesis is that the chance of a person having a heart attack is independent of whether or not the person smokes.

48. Consider conducting a chi-square test on the data from Question 46. Under the null hypothesis,
   (a) what is the estimate of the proportion of smokers who will have heart attacks?
   (b) what is the expected cell count for the "yes/yes" cell?

➤ **Solution**
   (a) We would estimate the proportion by using the marginal totals, so we get $\dfrac{59}{216} = .273$.

   (b) Taking (row total) × (column total)/(grand total) gives us $\dfrac{128 \times 59}{216}$ = 34.96. Note that this is the answer from part (a) of .273 multiplied by the number of smokers, 128.

49. Conduct a chi-square test on the data from Question 46. If α = .10, do you reject H_0?

➤ **Solution**

The chi-square value is 9.74, which exceeds the .05 critical value of 3.84 for a chi-square with 1 degree of freedom, so the p-value is less than .05 and we reject H_0.

---

50. A sample of 120 college students were asked if they had ever taken an AIDS test. (Assume that they answered truthfully.) Here are the data.

	Yes	No	Total
Male	8	51	59
Female	9	52	61
Total	17	103	120

(a) What proportion of the men have been tested?
(b) Of the people who have been tested, what proportion are men?

Answer (a) $\frac{8}{59} = .136$ (b) $\frac{8}{17} = .471$

---

51. Consider the data from Question 50. Under the assumption of independence between the rows and columns, what is the expected value of the "female/no" cell?

Answer $\dfrac{61 \times 103}{120} = 53.36$

---

52. Conduct a test of the hypothesis that the row and column variables in Question 50 are independent.

➤ **Solution**

The chi-square statistic is .036, which is very small, so the p-value is quite large and we do not reject H_0.

---

53. A survey organization polled a sample of Americans and asked them whether or not they approve of education "vouchers" that allow students in grades K to 12 to attend private schools at public expense. Two other organizations conducted similar surveys, but used different wording of the survey question. A social scientist wants to compare the percentage of people who supported the voucher idea in the three surveys.

(a) What kind of statistical test would one use with such data?

(b) How many degrees of freedom would the test statistic have?

➤ *Solution*

(a) These are categorical data that would be organized into a 3 × 2 table, with the 3 rows representing the different surveys and the 2 columns representing the yes/no response. Thus, a chi-square test could be done.

(b) The chi-square statistic would have 2 degrees of freedom.

MISCELLANEOUS REVIEW PROBLEMS

54. *Sleeping shrews.* The heart rates of sleeping shrews given in Question 11 of Topic 5 has the structure of a complete block design, with shrews as blocks and kinds of sleep as the factor of interest. Look over the data set. For which kind of sleep do the shrews' hearts beat fastest? Slowest? Which shrew had the fastest heart? The slowest? Is the variability greater from one shrew to the next, or from one kind of sleep to the next?

Answer

Hearts beat fastest during light slow-wave sleep, and slowest during deep slow-wave sleep. Shrew V had the fastest hear rate, and shrew I had the slowest. The variability from one shrew to another is quite a bit larger than the variability from one sleep phase to another.

55. *Confidence interval for predicted values: numerical drill.* The width of the confidence interval depends on the x-value for which you want a fitted y-value: the farther x is from the average of the observed xs, the wider the interval will be. Using data set (b) from Topic 5, Question 8, find 95% confidence intervals for fitted y-values corresponding to the x-values shown below, and use your intervals to complete the table; then plot "lower vs. x" and "upper vs x" on a set of xy-axes.

CONFIDENCE LIMITS

x	$x - x_{ave}$	Lower	Upper	Interval Width
0				
1				
2				
3				
4				

➤ **Solution**

The least squares line has equation $y = 2 + x/4$, with $SS_{Res} = 0^2 + (-1)^2 + (1)^2 = 2$, $df_{Res} = 3 - 2 = 1$, and $MS_{Res} = 2$.

The mean value for x is $x_{ave} = \frac{8}{3}$, and $S_{xx} = \sum (x - x_{ave})^2 = \frac{32}{3}$.

Confidence intervals for the fitted y-values are given by

$$\hat{y} \pm t \times SE, \text{ or } (2 + \frac{x}{4}) \pm (12.706)\sqrt{2}\sqrt{\frac{1}{3} + \frac{(x - x_{ave})^2}{(\frac{32}{3})}}$$

CONFIDENCE LIMITS

x	$x - x_{ave}$	Lower	Upper	Interval Width
0	−2.67	−16.64	19.30	35.94
1	−1.67	−12.26	15.43	27.69
2	−0.67	−9.17	12.84	22.01
3	.33	−8.45	12.62	21.07
4	1.33	−10.37	15.04	25.41
5	2.33	−13.92	19.09	33.01

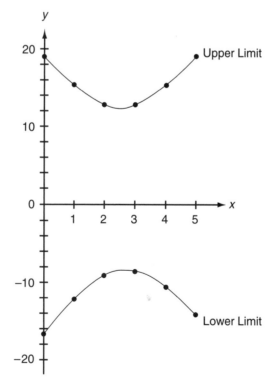

The goal of this experiment was to compare four seven-week training programs for infants, to see whether special exercises could speed up the process of learning to walk. One of the four training programs was of particular interest; it involved a daily twelve-minute program of special walking and placing exercises. The other three programs were different forms of controls. The second program, "exercise/control", involved daily exercise for twelve minutes, but without the special exercises. The third and fourth programs both involved no exercise (parents in these groups were given no instructions about exercise), but differed in follow-up; infants in the third program were checked every week, while those in the fourth group were checked only at the end of the study.

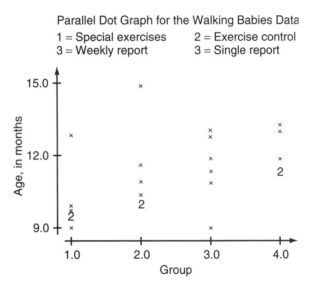

56. *Walking babies.* In the parallel dot graph above, identify three outliers, and notice that if you simply leave out all three outliers, you get a balanced design. Consider a new analysis that leaves out these three observations.

The data, including outliers, are given below:

	Treatment Groups						Averages
Special exercises	9	9.5	9.75	10	13	9.5	10.125
Control 1: exercise	11	10	10	11.8	10.5	15	11.375
Control 2: weekly report	11	12	9	11.5	13.3	13	11.675
Control 3: single report	13.3	11.5	12	13.5	11.5		12.350

Answer

The three values farthest from their group centers are the 13 in Group 1, and 15 in Group 2, and the 9 in Group 3. With these three values excluded, the design is balanced, with 5 observations in each group.

57. True or false. If you leave out all three outliers, then
   (a) SS_{Cond} will be larger than before;
   (b) df_{Res} will go from 19 down to 16;
   (c) MS_{Res} will get bigger, because df_{Res} gets smaller.

Answer
   (a) True: SS_{Cond} actually goes up, because the group means are farther apart when you exclude the outliers.
   (b) True.
   (c) False: The decrease in SS_{Res} more than offsets the decrease in df_{Res}, so that MS_{Res} goes down.
   Here's a comparison:

	Original Data	Outliers Excluded
SS_{Cond}	14.45	26.24
SS_{Res}	44.10	10.15

58. Compute new condition averages and condition effects. Then fill in the blanks in the following sentence four times, once for each condition: On average, babies in the _____ group walked _____ months _____ (sooner/later) than the overall average time.

Answer

Here are the new condition averages:

Group	Average Time (months)	Answer
1 (Special Exercise)	9.55	1.625 months sooner
2 (Exercise Control)	10.65	0.525 months sooner
3 (Control, Weekly Report)	12.15	0.975 months later
4 (Control, Final Report)	12.35	1.175 months later
Grand Average	11.175	

On average, babies in the first group walked 1.6 months sooner than the the overall average time.
On average, babies in the second group walked 0.5 months sooner than the the overall average time.

On average, babies in the third group walked 1.0 months later than the the overall average time.
On average, babies in the fourth group walked 1.2 months sooner than the the overall average time.

59. Now find MS_{Cond}, MS_{Res}, and the *F*-ratio. Taking the *F*-ratio at its face value, what do you conclude about the effects of the treatments?

Answer

MS_{Cond} = 8.75
MS_{Res} = 0.63
F = 13.79

Taking the *F*-ratio at face value, we would conclude that the treatments make a difference. It is important to emphasize that the conclusion here is valid *only* if you can justify excluding the three outliers.

The sizes of rectangles: interaction depends on the scale. Consider an experiment to study the effect of two factors, length and width, on the areas of rectangles. Suppose you choose three levels for the length factor (1", 4", 10") and two levels for the width factor (1", 4"). Take as your response the area of the rectangle, in square inches.

60. **(a)** Write out a table showing the factorial crossing. Put a suitable average response value in each of the six cells of your table.
(b) For each level of the length factor, measure the effect of width by computing the difference in areas for the two widths. Are these differences the same for all levels of the length factor?
(c) Draw and label an interaction graph.

Answer
(a)

	Length		
	1"	4"	10"
Width			
1"	1	4	10
4"	4	16	40

(b) Differences "due to" width are 3", 12", and 30", which are not equal: interaction is present.

(c)   **(a)**   **(b)**

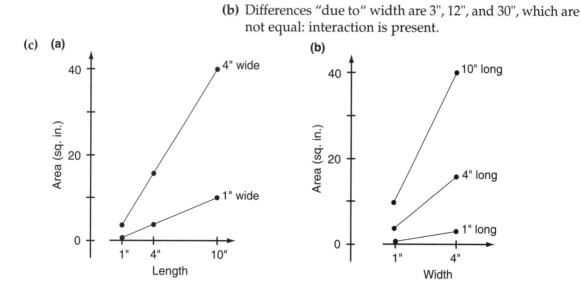

61. Now transform your data to logarithms, and repeat parts (a)–(c) from Question 60 in this new scale. Then (d) compare your two sets of results.

Answer

(a) Using logs to base 10, the transformed areas are:

		LENGTH	
	1"	**4"**	**10"**
Width			
1"	0	.60	1.00
4"	.60	1.20	1.60

(b) In the log scale, the differences "due to" width are all three equal to 0.60: there is no interaction.

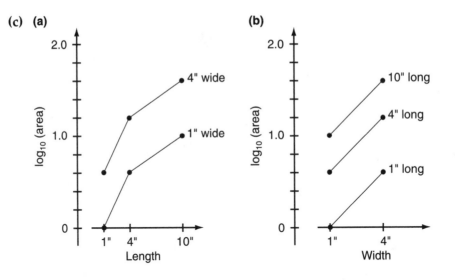

(d) In the original scale (area), differences due to width are different for different lengths: interaction is present. After transforming, in the new, logarithmic scale, the differences due to width are all the same: there is no interaction.

62. *Puzzled kids: decomposing a data set.* The data set below comes from a much larger study done at Mount Holyoke College in the Department of Psychology and Education. The response is the number of simple puzzles solved by a child in a fixed length of time. The data are for twelve kids, cross-classified by age and sex.

	AGE (YEARS)					
Sex	3		5		7	
Male	24	20*	18	16	31	53
Female	20	8	50	36	54	66

Adjusted slightly to simplify arithmetic.

(a) Construct the table of cell averages and draw the interaction graph.

(b) Which cell average seems most "out of line" with the others?

(c) What is your reason for judging it "out of line"?

(d) Fill in the blanks with either "boys" or "girls": The numbers suggest that at age 3, _____ are better than _____ at solving the kind of puzzles used in the experiment, but that within a few years, the _____ have overtaken the _____.

Answer

(a)

AGE (YEARS)

Sex	3	5	7
Male	22	17	42
Female	14	43	60

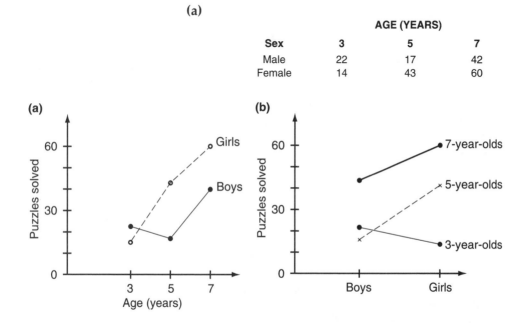

(b) The cell for the 5-year-old boys is most out of line

(c) Taken at face value, the averages say that 3-year-old boys are better at solving puzzles than 5-year-old boys.

(d) At age 3, *boys* are better than *girls* . . . within a few years, the *girls* have overtaken the *boys*.

63. *Compensating for parsnip webworms.* Under the control conditions of this study, wild parsnip plants averaged about a thousand seeds from their first set of flowers (primary umbels), about twice that many from the second set of flowers, but only about 250 from the third set, because the flowers in the third set tend to abort before the seeds are fully formed. However, for plants attacked by the parsnip webworm, which destroyed most of the primary umbels, the pattern was quite different: the seed production for primary, secondary, and tertiary umbels averaged about 200, 2400, and 1300, respectively.

(a) Put the cell averages in a two-way table, and label it.

(b) Draw both interaction graphs. Which one do you find easier to read? Why?

(c) Describe the interaction pattern in words, and discuss how the pattern illustrates the way the wild parsnips attacked by webworms compensated for the damage.

Answer

Attacked by webworms?	WHICH SET OF UMBLES?		
	Primary	Secondary	Tertiary
No	1000	2000	250
Yes	200	2400	1300

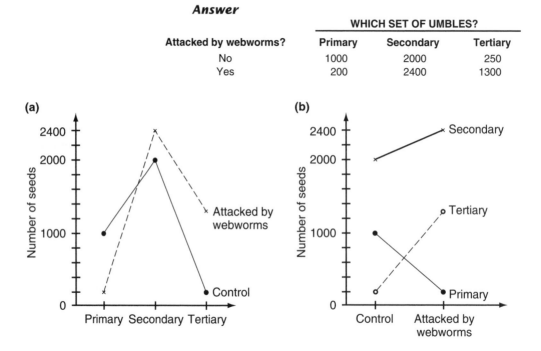

(c) For the control group, seed production in the primary and secondary umbels is successful, and so few seeds are produced in the tertiary umbels. For the plants attacked by webworms, the primary umbels produce few seeds. The secondary umbels compensate by producing a very large number of seeds, and the tertiary umbels produce almost half that many.

64. *Sugar metabolism.* The data below come from a two-way factorial experiment designed to study the effects of oxygen concentration on the amount of ethanol produced from two sugars, galactose and glucose, by *Streptococcus* bacteria. The response is the concentration of ethanol, in micromoles per .1 micrograms of sugar.

Sugar	OXYGEN CONC. (MICROMOLES)							
	0		**46**		**92**		**138**	
Galactose	59	30	44	18	22	23	12	13
Glucose	25	3	13	2	7	0	0	1

Do as complete an analysis as you can. Include parallel dot and interaction graphs, decomposition and ANOVA, and informal checks of the assumptions. If you decide the data should be transformed, your best bet is probably the log transformation, which is the one most often used for concentrations. (Note that because some of the concentrations are 0, you can't take their logs. In situations like this, standard practice is to add 1 to all the numbers first, then take logs.)

Answer

A plot of the data shows several important patterns:

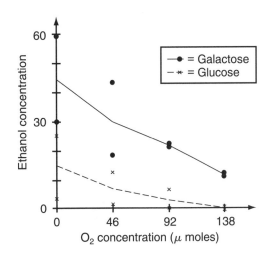

Ethanol production goes down as the oxygen concentration goes up, and the rate of decrease is greater at the lower concentrations of oxygen.

Ethanol production is much greater for galactose than for glucose, and the difference between the two sugars is greater at lower oxygen concentrations. In other words, the size of the difference is positively related to O_2 concentration.

Spread is positively related to the size of the observed values, with larger spreads many times bigger than the smaller spreads.

All three patterns suggest transforming. So does a plot of residuals versus fitted values:

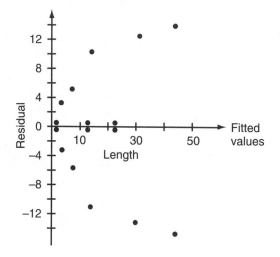

Here are the same data, reexpressed as $100 \log (1 \pm \text{concentration})$:

	O_2conc.			
	0	**46**	**92**	**138**
Galactose	178, 149	165, 128	136, 138	111, 115
Glucose	141, 60	115, 48	90, 0	0, 30

After transforming, the relation between ethanol production and oxygen concentration is roughly linear. Production is higher for galactose than glucose, and the rate of decrease, as O_2 concentration increases, is somewhat lower for galactose. However, the observed interaction effect is less pronounced than in the original scale. Although the within-group variation is not uniform, there is no longer (in the new scale) such a strong relationship between average and spread. A plot of residuals versus fitted values (not shown) suggests an oval balloon.

Here are the ANOVA tables

For ethanol concentrations:

Source	SS	df	MS	F	Table Value
Gr. Ave	4624.0	1			
Sugar	1806.25	1	1806.25	13.293	5.32
O_2 Conc.	1125.5	3	375.167	2.76	4.07
Inter.	118.25	3	39.4167	0.29	4.07
Residual	1087.0	8	135.875		
TOTAL	88240.0	16			

For $100 \times \log (1 + \text{conc.})$

Source	SS	df	MS	F	Table Value
Gr. Ave.	160,959.0	1			
Sugar	25,273.7	1	25,273.7	18.065	5.32
O_2 Conc.	10,398.3	3	3,466.1	2.478	4.07
Inter.	897.9	3	299.3	0.214	4.07
Residual	11,192.2	8	1399.025		
TOTAL	208,801.1	16			

In either scale, the conclusions are the same: only the difference between the two sugars registers as too big to be due to chance variability. The data give some reason to think that there may be an effect due to O_2 concentration, but the size of the effect is not large in comparison to chance error size. (A more sharply focused test for a linear trend in the O_2 effect would be likely, I think, to register as significant.) There is no evidence of an interaction between sugar and O_2 concentration.

65. Sabina Blaskovic (Oberlin College '97) collected data on whether or not students get 8 hours of sleep per night and what their majors are. The following table summarizes the data.

	MAJOR			
8 Hours of Sleep?	Humanities	Social Sciences	Natural Sciences	Total
Yes	4	5	4	13
No	11	17	11	39
Total	15	22	15	52

(a) State in words the null hypothesis that one would generally test with such data.

(b) How many degrees of freedom does the test statistic have?

(c) What is the expected cell count, under the null hypothesis, of the Humanities/Yes cell?

(d) The value of the test statistic is .11. (You do not need to check this calculation.) If $\alpha = .05$, do you reject H_0? Why or why not?

Answer

(a) The null hypothesis is that the probability that a student gets 8 hours of sleep is independent of the student's major.

(b) The test statistic has 2 degrees of freedom.

(c) The expected count is given by (the row total) × (the column total)/(the grand total), which is 39×15/52 or 11.25.

(d) For a chi-square statistic with 2 degrees of freedom, a value of .11 is quite small. The p-value is very large, so we do not reject H_0. (To get a p-value less than .05, so that we would reject H_0, we would have to have a test statistic value greater than 5.99.)

66. Researchers measured the pH level of precipitation in Colorado over $n = 150$ weeks, starting in 1975. They fit the model $Y_i = b_0 + b_1 X_i + e_i$ to the data, where $Y = $ pH of precipitation and $X = $ number of weeks after 1/1/75. The fitted model, with standard errors of the fitted parameters given in parentheses, is

$$\hat{Y} = 5.43 - .0053x.$$
$$SE \quad (.11) \ (.0013)$$

(a) Predict the pH level of precipitation on 1/1/76.

(b) Interpret the value .0053 in the context of this problem.

(c) Find a 95% confidence interval for b_1.

(d) The correlation coefficient is −.45. Suppose that X were recorded in years rather than in weeks. How would this affect the correlation coefficient? Explain your answer.

Answer

(a) $X=52$, so the predicted value is $5.43 - .0053 \times 52 = 5.1544$.

(b) On average, the pH level drops by .0053 with each passing week.

(c) Using 100 as the degrees of freedom for the t-multiplier, we have $-.0053 \pm 1.984(.0013)$ or $-.0053 \pm .00258$, which is $(-.00788, -.00272)$.

(d) The correlation coefficient does not depend on the units of measurement, so r would still be $-.45$.

67. (*From Australian Wildlife Research* (1984), via *Intro. to Probabilities and Statistics*, 8th ed., by Mendenhall and Beaver.) Mercury concentrations, in mg/g, were measured in the wing muscles of samples of three species of waterfowl. Here is a partial ANOVA table summarizing the data.

Source	df	SS	MS	F
Between Groups		0.0045		
Error		0.0655		
Total	26	0.0700		

(a) State H_0 in words.
(b) If $\alpha = .05$, do you reject the null hypothesis that there is no difference between the averages for the three species? Why or why not? (Give bounds on the p-value.)
(c) The average value of the 6 observations of the species Shelduck was .09. Construct a 95% confidence interval for μ, the population average.

Answer

(a) The null hypothesis is that the population average mercury concentrations are the same for all three species of waterfowl.
(b) The between groups mean square is $.0045/2 = .00225$. The within groups mean square is $.0655/24 = .00273$. Thus, $F = .00225/.00273 = .82$. The F statistic has 2 and 24 degrees of freedom and the p-value is quite large (larger than .10). Thus, we do not reject H_0 at the .05 level.
(c) The standard error is $\sqrt{.00273}/\sqrt{6} = .02133$. The t-multiplier has 24 degrees of freedom. Thus, the confidence interval is given by $.09 \pm 2.084(.02133)$ or $(.046, .134)$.

68. (Based on data in *Biometry* by Sokal and Rohlf.) Researchers wanted to determine whether or not a particular antiserum was effective, so they gave a dose of pathogenic bacteria to 57 mice and then gave those mice the antiserum; 13 of them died. They gave the bacteria but no antiserum to 54 other mice; 25 of them died. Here is the cross-classification table of the data.

	Died	Lived	Total
Bacteria and antiserum	13	44	57
Bacteria only	25	29	54
Total	38	73	111

For this group of 111 mice, is presence of the antiserum independent of outcome (lived versus died)? Why or why not?

Answer

The table of expected counts is given below.

	Died	Lived	Total
Bacteria and antiserum	19.51	37.49	57
Bacteria only	18.49	35.51	54
Total	38	73	111

The chi-square statistic has 1 degree of freedom and is given by $\dfrac{(13 - 19.51)^2}{19.51} + \dfrac{(44 - 37.49)^2}{37.49} + \dfrac{(25 - 18.49)^2}{18.49}$

$+ \dfrac{(29 - 35.51)^2}{35.51} = 6.788$. The p-value is between .001

and .01. This gives us a strong indication that the outcome depends on the presence of the antiserum. (If α is .01 or larger, we would reject H_0.)

69. (Based on an article in the *New England Journal of Medicine.* 315 (1986): 977-82; via Berry and Lindgren, *Statistics: Theory and Methods.*) Researchers collected data on coffee drinking and presence or absence of coronary heart disease (CHD) for a sample of 1040 persons. Here are the data:

	CUPS PER DAY				
	0	1–2	3–4	> 4	Total
CHD	4	17	17	9	47
No CHD	185	457	226	125	993
Total	189	474	243	134	1040

Suppose we wish to test the null hypothesis that CHD is independent of coffee-drinking habits.

(a) How many degrees of freedom does the test statistic have?

(b) What is the expected cell count, under the null hypothesis, of the CHD/1–2 cell?

(c) The value of the test statistic is 8.44. What are the bounds on the p-value?

(d) Suppose $\alpha=.05$. Do you reject H_0? Why or why not?

Answer

(a) The test statistic has 3 degrees of freedom.

(b) The expected count is given by (the row total) × (the column total)/(the grand total), which is $47 \times 474/1040$ or 21.42.

(c) The p-value is between .02 and .05.

(d) We reject H_0 since the p-value is less than .05.

70. (Data from R. Iman, *A Data-Based Approach to Statistics*.) Researchers conducted an experiment to compare four types of cigarette filters. The response variable measured was the amount of tar, in milligrams, that passed through the filter. Here is a partial ANOVA table summarizing the results.

Source	SS	df	MS	F
Between Groups	7.62			
Within Groups	19.60	30		
Total	27.22			

(a) How many Between Groups degrees of freedom do we have?

(b) How many cigarettes (observations) were in the study?

(c) We use ANOVA to test a null hypothesis, H_0. State H_0 in words in the context of this setting. Be specific.

(d) Find the value of the test statistic that is used to test H_0. (You do not need to complete the test.)

(e) What assumptions are necessary for the test to be valid?

Answer

(a) There are 3 between groups degrees of freedom, since there are 4 groups.

(b) There are 33 total degrees of freedom, so there were 34 cigarettes in the study.

(c) The null hypothesis is that the population average amount of tar that passes through a cigarette filter is the same for all four types of cigarettes.

(d) The between groups mean square is $7.62/3 = 2.54$. The within groups mean square is $19.60/30 = .6533$. Thus, $F = 2.54/.6533 = 3.89$.

(e) Effects should be constant and additive. Chance errors should be normal and independent, with the same SDs.

71. I had Data Desk construct parallel dot plots of the raw data from the preceding question. I then conducted a test of H_0: $\mu_1 = \mu_2 = \mu_3 = \mu_4$ and rejected H_0 at the $\alpha = .05$ level. However, when I tested H_0: $\mu_2 = \mu_3 = \mu_4$ using $\alpha = .05$, I did not reject H_0.

(a) Your job is to sketch a graph of the parallel dot plots of the data. That is, based on what I told you about the tests you should have an idea of how the data look. Use that idea to draw a graph. Indicate the sample means with horizontal bars.

(b) It is possible to get data with the same sample means that you graphed in part (a), but for which the hypothesis $H_0: \mu_1 = \mu_2 = \mu_3 = \mu_4$ is not rejected at the $\alpha = .05$ level? Provide a graph of this situation. That is, keep the same sample means (horizontal bars) you had from part (a), but show how the data would have been different if H_0 was not to be rejected.

Answer

(a) The last three types have similar means, but the mean of the first type differs from the others. The following plot shows one way this could happen.

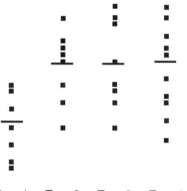

Type 1   Type 2   Type 3   Type 4

(b) If the sample means are to remain the same but H_0 is not to be rejected, then there must be more variability within the groups. The following plot shows this situation.

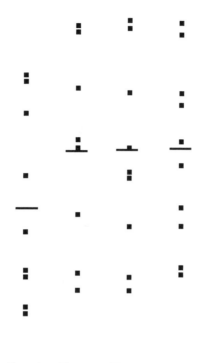

Type 1   Type 2   Type 3   Type 4

72. (Based on an article in the Cleveland *Plain Dealer* on 9/9/95.) A group of patients with ALS (Lou Gehrig's disease) were randomly assigned to get either Riluzole or a placebo. After 18 months some were still alive while others were not. Here are the data.

	Alive	Dead	Total
Riluzole	314	236	550
Placebo	275	275	550
Total	589	511	1100

Use these data to conduct a test of the null hypothesis that the chance of surviving at least 18 months is independent of treatment group (Riluzole versus placebo); use a directional alternative and use $\alpha = .01$. Provide all steps, including degrees of freedom, bounds on the *p*-value, and your conclusion regarding H_0.

Answer

The table of expected counts is given below.

	Alive	Dead	Total
Riluzole	294.5	255.5	550
Placebo	294.5	255.5	550
Total	589	511	1100

The chi-square statistic has 1 degree of freedom and is given by $\dfrac{(314 - 294.5)^2}{294.5} + \dfrac{(236 - 255.5)^2}{255.5} + \dfrac{(275 - 294.5)^2}{294.5} + \dfrac{(275 - 255.5)^2}{255.5}$ $= 5.56$. For the directional alternative, the p-value is between .005 and .01. We reject H_0 at the .01 level.

73. (From Cameron, E. and Pauling, L. (1978) "Supplemental ascorbate in the supportive treatment of cancer: re-evaluation of prolongation of survival times in terminal human cancer." Proceedings of the National Academy of Science, 75:4538-4542.) Patients with differing kinds of advanced cancers were treated with ascorbate. The purpose of the study was to determine if patient survival time differed with respect to the organ affected by the cancer. Here is a partial ANOVA table summarizing the results. (In this study, a "group" is a type of cancer.)

Source	SS	df	MS	F
Between Groups	3295.04	4		
Within Groups	7495.27	59		
Total	10790.31			

(a) How many types of cancer (groups) were in the study?
(b) How many patients were in the study?
(c) We use ANOVA to test a null hypothesis, H_0. State H_0 in symbols.
(d) Find the value of the test statistic that is used to test H_0 and the degrees of freedom for the test statistic. (You don't need to conduct the entire test).
(e) Jo says that if we want to find out if all means are equal we should just do a bunch of two-sample t-tests (from Topic 10) with $\alpha = .05$ each time, (to compare each pair of means) rather than an ANOVA F-test. What is your response to Jo? Explain your reasoning!

Answer

(a) There were 5 types of cancer in the study. (There are 4 between groups degrees of freedom, so there were 5 groups.)

(b) There were 64 patients in the study. (There are 63 total degrees of freedom.)

(c) $H_0: \mu_1 = \mu_2 = \mu_3 = \mu_4 = \mu_5$

(d) The between groups mean square is $3295.04/4 = 823.76$. The within groups mean square is $7495.27/59 = 127.038$. Thus, $F = 823.76/127.038 = 6.484$.

(e) If we do a series of t-tests, the probability of making at least one type I error somewhere along the way is quite a bit larger than .05 (i.e., the .05s sort of "add up" over the tests). By doing an analysis of variance we control the overall type I error rate.

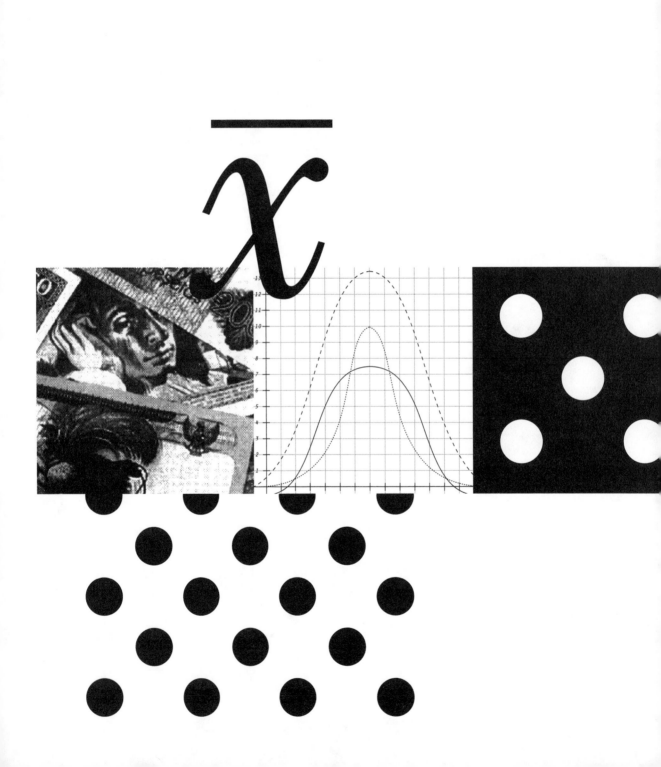

Nonparametric Methods

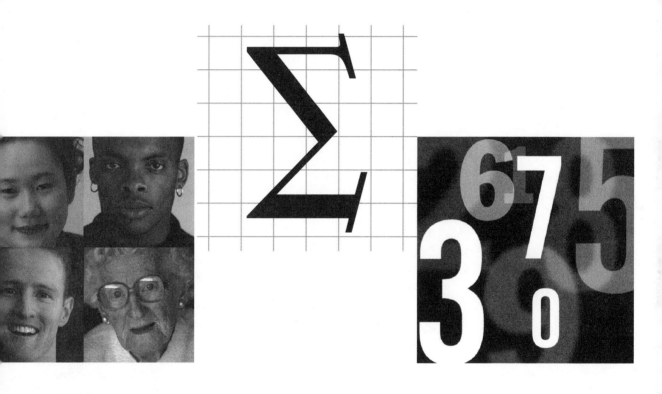

Summary

13.1 IN A NUTSHELL

Where does this topic fit in? Nonparametric methods offer you an alternative to the methods of inference summarized in Topics 10 and 12. If the shape of your distribution is not too far from normal, then you don't need special alternatives. On the other hand, if your distributions have long, heavy tails, with several extreme values, the methods of this chapter are more trustworthy than methods designed for normal-shaped data. (Yet a third possibility, in between the traditional and nonparametric methods, is to transform your data to ranks, then apply the traditional method to the ranks. Some statisticians recommend this in preference to using nonparametric methods.)

I've organized my summary into two parts: how to think about the ideas, and how to apply the methods.

What are the main ideas? There are essentially just two: the rank transformation, and the box model for the null hypothesis. All else is detail. The **rank transformation** replaces your data values with the rank orders. For example, the data 0, 2, 4, 6, 100 gets changed to 1, 2, 3, 4, 5. Notice how the 100—a clear outlier—gets changed to 5, which is the same distance from its neighboring value as the others are. This illustrates one reason why nonparametric methods can be useful for data from long-tailed distributions. (Tied values get the average of the ranks that would be assigned if they were not tied. So 0, 2, 2, 6, 100 becomes 1, 2.5, 2.5, 4, 5.) The **box model** for your **null hypothesis** depends on the particular methods, but the basic idea is this: you think of the ranks of your observations as being written on tickets, put into a box, and mixed thoroughly; you then imagine drawing out one set of tickets for each group (each treatment group, control group, etc.), and computing a summary number. The significance level (p-value) is the chance of getting a summary value as extreme as the one for the actual data.

How do you apply the methods? There are two main things to learn: how you find the p-value (i.e., the mechanics of the test), and when you should use each method. The best method for finding p-values is simple—get a computer to do it for you. Learning when to use a method is something I discuss in connection with the particular methods.

13.2 SIGNED RANK TEST (WILCOXON TEST) FOR ONE SAMPLE OR PAIRED DATA

When to use it The signed rank test is an alternative to the one-sample and paired t-tests (Topic 10) for heavy-tailed distributions. For moderate to large samples from long-tailed distributions, this test will be more powerful than the t-test. However, for normal-shaped data, it is less powerful than the t-test, and for skewed data, the computed p-values will not be correct.

The test statistic The null hypothesis is that the observed values come from a symmetric distribution with center at 0. The alternative hypotheses are that the values tend to be greater than 0 (right-sided alternative H_+), or less than 0 left-sided alternative H_-), or that one or the other of the one-sided alternatives is true ($H_{\neq}$). To compute the test statistic, first throw out any 0s. Ignore the signs (+ or −) and make a parallel stem plot of the non-zero data with the values that were initially positive in one group and the values that were initially negative in the other. Then assign one set of ranks to all the values, and sum the ranks in the two groups. The test statistics are:

$T = T_+$ = sum of the ranks for the values that were initially positive, used for testing H_0 versus H_+;

$T = T_-$ = sum of the ranks for the values that were initially negative, used for testing H_0 versus H_-; and

T = the smaller of T_+ and T_-, for testing H_0 versus $H_{\neq}$.

The box model For each rank, toss a fair coin to decide whether it gets assigned a + or −. There are $2 \times 2 \, x \ldots x^2 = 2^n$ ways to assign either a + or − to the set of ranks, and under the null hypothesis, these 2^n assignments are equally likely. The p-value is the chance of getting a value of T as large or larger than the one computed from the actual data.

A normal approximation If your sample size is 30 or more, you can standardize and use a z-test (Topic 9). To do this, use the following values for the mean and standard deviation.

$$\text{Mean } \mu = n(n + 1)/4 \quad \text{and} \quad \text{SD } \sigma = \sqrt{2(n + 1)\mu/6}$$

Correction for ties See page 125 of Eric Lehmann (1975), *Nonparametrics: Statistical Methods Based on Ranks*. San Francisco: Holden and Day.

13.3 RANK SUM TEST FOR TWO INDEPENDENT SAMPLES

When to use it The signed rank test is an alternative to the two-sample t-test (Topic 10) for skewed and/or heavy-tailed distributions. For moderate to large samples from skewed or long-tailed distributions, this test will be more powerful than the t-test, and the *p*-values will not be distorted by these features of the distribution. However, for normal-shaped data, it is less powerful than the *t*-test, and it is not suitable when the spreads for the two groups are very unequal. (Unequal spreads suggest transforming to a new scale.)

(WILCOXON TEST, EQUIVALENT TO MANN-WHITNEY TEST)

The test statistic The null hypothesis is that the two sets of observed values come from identical distributions. To compute the test statistic, first make a parallel stem plot of the data. Then assign one set of ranks to all values, and sum the ranks in the first group.

The box model The box contains tickets with the ranks on them. Randomly draw out one ticket for each observation in the first group, and add up these ranks. The *p*-value is the chance of getting a value more extreme than the value of the observed test statistic.

A normal approximation If both sample sizes are 10 or more, you can standardize and use a *z*-test (Topic 9). To do this, use the following values for the mean and standard deviation.

$$\text{mean } \mu = [\text{average rank}](\text{sample size}) = [(n_1 + n_2 + 1)/2](n_1)$$
$$\text{SD } \sigma = \sqrt{n_2 \mu / 6}$$

Correction for ties See page 20 of Eric Lehmann (1975), *Nonparametrics: Statistical Methods Based on Ranks*. San Francisco: Holden and Day.

13.4 KRUSKAL-WALLIS TEST FOR THREE OR MORE INDEPENDENT SAMPLES

When to use it The Kruskal-Wallis test is an alternative to one-way analysis of variance (Topic 12) for skewed and/or heavy-tailed distributions. For moderate to large samples from skewed or long-tailed distributions, this test will be more powerful than ordinary ANOVA, and the *p*-values will not be distorted by these features of the distribution. However, for normal-shaped data, it is less powerful than ANOVA, and it is not suitable when the spreads for the groups are very unequal. (Unequal spreads suggests transforming to a new scale.)

The test statistic The null hypothesis is that the groups of observed values all come from the same distribution. To compute the test statistic, first make a parallel stem plot of the data. Then assign one set of ranks to all the values, and sum the ranks in each group. Using T_i for the total , n_i for the size of group i, and N for the total sample size, the test statistic equals

$$[12/N(N+1)] \{(\sum T_i^2/n_i) - 3(N+1)\}.$$

(Unless there are ties, the test statistic is equivalent to the usual sum of squares for treatments (Topic 12), but it is computed from the ranks. In practice you should use a computer to get the value of the test statistic and the p-value.)

The box model The box contains tickets with the ranks on them. Randomly draw out one ticket for each observation in the first group, and add up these ranks to get T_1. (Don't replace these tickets.) Now draw one ticket out of the box for each observation in the second group, and add up these ranks to get T_2. Keep going in this way until you have one total for each group. Then combine the totals as above to form a value of the test statistic. The p-value is the chance of getting a value more extreme than the value of the observed test statistic.

Approximate p-value You can get an approximate p-value by assuming that when the null hypothesis is true, the sampling distribution of the test statistic is chi-square, with degrees of freedom equal to (# groups − 1).

Correction for ties Divide the value of the test statistic by $\{1 - \sum(d_1^3 - d_1)/(N^3 - N)\}$, where d_1 is the number of data values tied at rank = 1, and N is the total number of observations.

13.5 FRIEDMAN TEST FOR DATA FROM A RANDOMIZED COMPLETE BLOCK DESIGN

When to use it The Friedman test is an alternative to analysis of variance for a randomized complete block design (Topic 12) for skewed and/or heavy-tailed distributions. For moderate to large samples from such distributions, this test will be more powerful than ordinary ANOVA, and the p-values will not be distorted by the skewness, long tails, or outliers. However, for normal-shaped data, it is less powerful than ANOVA, and it is not suitable when the spreads for the groups are very unequal. (Unequal spreads suggest transforming to a new scale.)

The test statistic The null hypothesis is that within each block, all orderings are equally likely, that is, there is no treatment effect. To compute the test statistic, first arrange the data in a rectangle with blocks as rows and treatments as columns. Transform to ranks separately within each row. Unless there are ties, the test statistic is equivalent to the usual sum of squares for treatments (Topic 12), but it is computed from the ranks. In practice you should use a computer to get the value of the test statistic and the p-value. Using T_i for the total for the ith treatment, t for the number of treatments, and b for the number of blocks, the test statistic equals

$$[12/bt(t+1)]\sum(T_i - b(t+1)/2)^2.$$

The box model Under the null hypothesis, all possible orderings within each block are equally likely. This means you can create random data sets one block at a time, by drawing at random from a box of tickets numbered $1, 2, \ldots t$, where t is the number of treatments. Drawing out the tickets one at a time in random order gives you one block's worth of ranks. Repeat the process for each block, and then compute the value of the test statistic for the random data. The p-value is the chance of getting a value of the statistic at least as large as the one based on your actual data.

Approximate p-value You can get an approximate p-value by assuming that when the null hypothesis is true, the sampling distribution of the test statistic is chi-square, with degrees of freedom equal to (# treatments − 1).

Correction for ties Divide the value of the test statistic by $\{1 - (\sum\sum d_{i1}^3 - bt)/[bt(t^2 - 1)]\}$, where d_{i1} is the number of data values in block i tied at rank = 1.

13.6 SIGN TEST FOR PAIRED DATA

When to use it Just as paired data is a special case of data from a complete block design, with pairs as blocks, the sign test is equivalent to the Friedman test for blocks of size 2. So the sign test is an alternative to the paired t-test, for situations where the distribution of differences is not roughly normal shaped.

The test statistic Question 20 shows how to do the sign test.

13.7 RANK CORRELATION FOR PAIRED DATA

When to use it The rank correlation is an alternative to the usual correlation coefficient for paired data (Topics 5 and 12). Whereas the usual (Pearson's) correlation measures the strength of *linear* association, the rank (Spearman's) correlation measures *monotonic* association: To what extent does y tend to increase when x increases (regardless of whether the pattern of increase is linear)? Or, to what extent does y tend to decrease as x increases?

How to compute it Transform the values of x to ranks, then do the same for the values of y, and then compute the usual correlation coefficient using the ranks.

13.8 RUNS TEST FOR SEQUENCE DATA

When to use it If you have a sequence of observations and you want to test the null hypothesis that the order is random, you can use a runs test.

Computing the test statistic Choose a threshold value K, such as the median of the observations. Replace data values $\leq K$ with 0, values $>K$ with 1, so that your sequence is formed just of stretches of 0s and stretches of 1s. Each of these stretches is called a run. The test statistic is the number of runs.

The box model According to the null hypothesis, all rearrangements of the sequence of 0s and 1s are equally likely. Set up a box with one ticket for each 0 and 1 in your sequence, draw out tickets randomly one at a time, and count the number of runs in the sequence you get. The p-value is the chance of getting a value more extreme than the value you got from the actual data.

A normal approximation If both the number of 0s and the number of 1s are at least 20, you can get an approximate p-value by standardizing and using a z-test (Topic 9). To standardize, use the following mean and standard deviation, where n is the sample size and p is the proportion of data values $\leq K$.

$$\text{mean } \mu = 2np(1 - p) + 1 \text{ and } SD \; \sigma = \sqrt{\{2p(1 - p)/n\}[2p(1 - p) - 1}$$

If you use the book by Triola, you can compute the mean and standard deviation in terms of the numbers n_1 and n_2 of observation $\leq K$ and $>K$.

$$\text{mean } \mu = 2n_1 n_2/(n_1 + n_2) + 1 \text{ and } SD \; \sigma = \sqrt{\frac{2n_1 n_2(2n_1 n_2 - n_1 - n_2)}{(n_1 + n_2)^2(n_1 + n_2 - 1)}}$$

Self-Testing Questions

1. The sum of the ranks of 6 observations is 21. Another observation is added. What is the resulting sum of the ranks?

➤ **Solution**

28. The sum of the ranks of a set of observations is determined by the sample size. Adding another observation just adds that rank to the final sum.

2. In a ranking, the smallest observation has rank 3. How many of the observations are tied for the first rank?

➤ **Solution**

All of the observations that are tied have the same rank. There must be a five-way tie, since $(1 + 2 + 3 + 4 + 5)/5 = 3$.

3. A set of data is ranked. After a log transformation of the data, the observations are again ranked, and these ranks are subtracted from the ranks of the original untransformed data. What is the sum of the differences?

➤ **Solution**

The log transformation (assuming there are no rounding errors) keeps observations in the same order as they were before transformation. Thus, each difference is 0.

4. Consider the following data for motor vehicle accidents by region of the country for 1980 and 1985.

Region	1980	1985	Difference	Rank
New England	2246	1873	373	2
MidAtlantic	6009	4924	1085	7
East Northern Central	8000	6602	1398	9
West Northern Central	4244	3158	1266	8
Southern Atlantic	9767	9389	378	3
East Southern Central	3995	3664	331	1
West Southern Central	7252	6197	1061	6
Mountain	4071	3409	662	4
Pacific	7892	6943	949	5

If the 1985 Southern Atlantic value was found to be 9000 rather than 9389, would the p-value for the signed rank test change?

➤ *Solution*

The p-value won't change. All of the differences remain positive, so the test statistic and p-value do not change.

5. In the accident data from Question 4, what would be the effect on the p-value for the signed rank test if the 1985 value for the Mountain region changed to 5000?

➤ *Solution*

The p-value would increase since the Mountain difference would be negative.

6. In a roughly symmetric distribution, what two factors commonly suggest that a non-parametric test (Kruskal-Wallis) would be a better choice than the usual method (ANOVA and F-test), which assumes the data are roughly normal?

Answer Outliers and heavy tails

7. In a ranked sum test of 8 observations, the one-sided test statistics, T_+ and T_- are equal, and the ranks for the 4 observations in the control group are tied. What is the rank of all these control group observations?

➤ *Solution*

The ranks must lie in the middle of the distribution of ranks. Thus, if they were not tied, they would be 3, 4, 5, and 6. However, since they are tied, they all have rank 4.5.

8. A team of biologists counted the numbers of species of fish in each of 6 locations along a river. At each location, they found the number of species in a pool of water and the number in a nearby riffle. Here are the data.

Location	Pool	Riffle
1	6	3
2	6	3
3	8	4
4	1	2
5	4	3
6	6	2

If we conduct a sign test of these data, what is the value of the test statistic?

➤ **Solution**

There are 5 positive differences, so the test statistic's value is 5.

9. Researchers asked several smokers how many cigarettes they had smoked the previous day. There were 4 women and 6 men in the sample. Here are the data.

Women	Men
4	2
7	2
20	5
20	6
	8
	16

Consider a rank sum test of the hypothesis that there is no difference between men and women. Find the rank sum for the women.

➤ **Solution**

The ranks for the women are 3, 6, 9.5, and 9.5, so the rank sum is 28.

10. Consider the data from Question 9. Could we conduct a sign test here?

➤ **Solution**

No, we do not have paired data.

11. When the distribution of differences is skewed, which is better: the Friedman test (equivalent to the sign test) or the Wilcoxon test?

Answer The Friedman test

12. An experimenter decides that the signed rank and Friedman tests are the most appropriate non-parametric tests for her data. What does this say about the nature of the data?

Answer They are paired data.

13. In a Wilcoxon rank sum test on a data set with lots of tied values, what will be the effect of the ties on the *EV* and *SE*?

Answer

The *SE* will be decreased, but the *EV* will be unchanged.

14. In a paired data sample where the differences are symmetric about 0, what will be the effect on the power of the signed rank test if there are a large number of pairs with difference = 0?

➤ ***Solution***

The power will decrease, since the signed rank test discards the tied pairs, which reduces the effective sample size. (The *t*-test keeps the 0s, and they lower the SD.)

15. If you have data pairs (x, y) and the distribution of $x - y$ is highly skewed, is the signed rank test a good choice for testing whether the mean of the differences have mean 0?

➤ ***Solution***

No. The signed rank test is not appropriate for skewed distributions.

16. For the accident data from Question 4, the 9 regions are divided into north and south with 5 regions and 4 regions, respectively. If the null hypothesis is true for the rank sum test, all possible ways of sorting the ranks into two groups are equally likely. Are there any restrictions on the sizes of the groups?

Answer

The group sizes must be 5 and 4, as in the actual data.

17. Consider the data from Question 8. Could we conduct a rank sum test within this situation?

➤ **Solution**

No, the data are paired, so the rank sum test would not be appropriate.

18. A control chart for a process showed that sometimes the reading was above the target (A) and sometimes the reading was below the target (B). Here are the data: AABAB-BAAABBABBBA. How many runs were there?

Answer

There were 9 runs, with run lengths of 2, 1, 1, 2, 3, 2, 1, 3, and 1.

19. Two people were asked to independently rate the quality of some popular candy, using a scale of 1 to 100. Here are the ratings they came up with.

Candy	Rating 1	Rating 2
Snicker's	95	90
Milky Way	75	84
M & M's	100	95
Kit Kat	70	72
Nestle's Crunch	84	100
Twix	56	88

Find the value of the rank correlation between the two sets of ratings.

➤ *Solution*

The ranks and the differences in ranks are as follows.

Candy	Rank 1	Rank 2	Difference
Snicker's	2	3	−1
Milky Way	4	5	−1
M & M's	1	2	−1
Kit Kat	5	6	−1
Nestle's Crunch	3	1	2
Twix	6	4	2

The sum of the squared differences is 12. Thus, the rank correlation coefficient is $1 - \dfrac{6 \times 12}{6 \times (35)} = 1 - .343 = .657$.

20. *Cool mice: the sign test.* The cooling constant measures the rate at which the temperature of a warm object returns to the temperature of its environment. Each kind of object has its own cooling constant, and medical examiners rely on this fact when they use the temperature of a corpse to determine the time of death.

Imagine for a moment that you have decided to commit murder. To cook up an alibi for yourself, you plan to heat up your victim's body to mislead the coroner about the time of death. Do reheated bodies cool at the same rate as freshly killed ones?

Believe it or not, in 1951 J. S. Hart published (in the *Canadian Journal of Zoology*) the results of a study designed to answer this question. He chose for his response the cooling constant; he used 19 mice as his experimental material; and there were two conditions of interest: freshly killed and reheated. Hart used a complete block design, with each mouse as a block. Each mouse provided two observational units: each mouse was measured twice, once when freshly killed, and then again after reheating.

Hart's Cooling Constants for Mice under Two Conditions

Mouse (Block)	Freshly Killed	Reheated
1	573	481
2	482	343
3	377	383
4	390	380
5	535	454
6	414	425
7	438	393
8	410	435
9	418	422
10	368	346
11	445	443
12	383	342
13	391	378
14	410	402
15	433	400
16	405	360
17	340	373
18	328	373
19	400	412

Suppose you want to test the null hypothesis that for each mouse, there is a 50-50 chance that the cooling constant for fresh will be greater than the cooling constant for reheated, just as there is a 50-50 chance that a fair coin lands heads.

(a) Think of your null hypothesis as describing a chance mechanism: if the hypothesis is true, then each mouse corresponds to a random draw (with replacement) from a box with two tickets. The mouse experiment corresponds to how many draws from the box?

(b) Take as your summary statistic the numbers of times Fresh is higher than Reheated.
  (i) What is the observed value from the data?
  (ii) What is the *EV* for the summary statistic?

(c) Here is a smooth approximate version of the probability histogram for the summary statistic. Guess the SD.

Fresh is higher	Reheated is higher

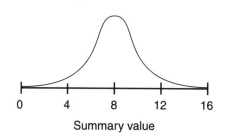

Summary value

(d) Suppose you want to use your summary statistic to test your null hypothesis. You might reason as follows: "If my hypothesis is true, then the histogram gives the sampling distribution of the summary statistic. Outcomes near the center (*EV*) are likely, and give me no reason to suspect the hypothesis is false. On the other hand, outcomes in the tails, far away from the *EV* on either side, are unlikely (unless my hypothesis is wrong).

The 2.5th percentile for the distribution is about 5.2, and the 97.5th percentile is about 13.8. Explain how to use these numbers to complete a test of the null hypothesis. Then do the test and state your conclusion.

(e) Notice that you could do an equivalent test by choosing as your test statistic:

Absolute value of (# times Fresh higher – # times Reheated higher).

Explain how you could use simulation to find the 5% critical value.

Answer

(a) 19 draws, one for each mouse.
(b) **(i)** The observed value is 12: there were 12 mice with the Fresh reading higher than Reheated.
(ii) The null hypothesis is that $Pr(\text{Fresh} > \text{Reheated}) =$

   0.5. So $EV = np = (19)(.5) = 9.5$

(c) Use the fact that the flex points (where the curvature changes) are above the values $\mu - \sigma$ and $\mu + \sigma$. The flex points are roughly 4 <u>units apart</u>, making the SD about 2. (The exact value is $\sqrt{np(1 - p)} = 2.18$.)
(d) A 5% fixed-level test of H_0: $p = \frac{1}{2}$ versus H_A: $p \neq \frac{1}{2}$ will reject H_0 if the observed value in part (b) is less than 5.2 or greater than 13.8. Since the observed value of 12 falls between 5.2 and 13.8, we conclude that the evidence against H_0 is not strong enough to reject it.
(e) Use sets of 19 draws from the box model in part (a) for your simulation. For each set of 19 draws, compute the value of the test statistic, and keep track of the values for, say, 1,000 sets of 19 draws. Use these results to find the value exceeded by 50 (5%) of the 1,000 values.

21. (Based on an article in the *Journal of Reproduction and Fertility* (1984): 385–93; via Devore and Peck, *Statistics*.) Seven rams were used in a study of the effect of stress on the release of luteinizing hormone (LH). The following table gives LH release rates (in ng/min) before and after treatment with the drug adrenocorticotrophin.

Ram	1	2	3	4	5	6	7
Before	2400	1400	1375	1325	1200	1150	850
After	250	1425	1100	800	850	925	700
Difference	2150	−25	275	525	350	225	150

(a) Using $\alpha = .10$, conduct a sign test to determine whether treatment increases LH release rate. Provide all steps, including bounds on the *p*-value and conclusion.

(b) What assumptions are necessary for the sign test to be valid?

Answer

(a) The null hypothesis for the sign test is that there is a 50-50 chance of any particular difference being positive. We have 6 positive differences out of 7. The probability of getting at least 6 out of 7 positive differences if H_0 is true is $7 \times (.5^6) \times (.5) + .5^7$, which is $.0547 + .0078 = .0625$.

(b) The observations must be a random sample from the population.

Data Sources

A. DATA SOURCES FOR THE REVIEW MATERIAL*

Abbreviations

AA: *American Almanac.* Austin, TX: The Reference Press. Unless a year is given, the edition is 1992–1993. This almanac is based on the *Statistical Abstract of the United States.* Washington, DC: United States Bureau of the Census.

ABC: *America's Best Colleges 1996* : Washington, D.C.: US News & World Report.

WABF : Farmighetti, Robert, ed. *World Almanac and Book of Facts.* Mahwah, NJ: Funk & Wagnalls.

Topic 1

Witches. Boyer, P. and S. Nissenbaum (1972). *Salem-Village Witchcraft: A Documentary Record of Local Conflict in Colonial New England.* California: Wadsworth Publishing Company.

Finger tapping. Scott, C. C. and K. K. Chen (1944). "Comparison of the action of 1-ethyl theobromine and caffeine in animals and man," *J. Pharmacol. Exptl. Therap.* **1**, 89–97.

Topic 2

Diet motivators. *USA Today*, 8/14/95, p. 1.

Buying books. *Daily Hampshire Gazette*, 11/8/95, p. 18.

Nuclear energy. *WABF 1994*, p. 156.

Brain weights, body weights. Crile, G. and D. P. Quiring (1940). "A Record of the Body Weight and Certain Organ and Gland Weights of 3690 Animals," *Ohio J. of Science*, **5**: 219–259.

Old Faithful. Weisberg, Sanford (1985). *Applied Linear Regression*, 2nd ed. New York: John Wiley and Sons, p. 230.

English monarchs. Crystal, David, ed. (1994). *The Cambridge Factfinder.* Cambridge: Cambridge University Press, pp. 174–175.

Retail establishments. *AA*, p. 765.

Hazardous waste sites. *AA*, p. 215.

Topic 4

Life expectancy. *Daily Hampshire Gazette*, 12/26/95

Microwaves. *WABF 1996*, p. 203.

---

*Sources that are listed in the text of the questions are not repeated here.

Sunspots. Sunspot Index Data Center, Royal Observatory of Belgium, http://www.oma.be/KSB-ORB/SIDC/

Faculty salaries. *The NEA 1995 Almanac of Higher Education*, p. 9.

Topic 5

Mental activity. Finney, D. J. (1946). *Biometrics Bull.*, **2**, 53, via Snedecor and Cochran, p. 426.

Anscombe data. Anscombe, F. J. (1973). "Graphs in statistical analysis," *Amer. Statist.*, **27**, 17–21.

Urban population. *AA*, p. 27.

Dormitory population. *AA*, p. 57.

Player costs and total revenues. *Financial World*, 5/20/96, p. 56–63.

SAT scores. *WABF 1994*, p. 199.

Dopamine and schizophrenia. Sternberg, D. E., D. P. Van Kammen and W. E. Bunney (1982). "Schizophrenia: dopamine b-hydroxylase activity and treatment response," *Science*, **216**, 1423–1425.

Leafhopper survival. Allen, D. M. and F. B. Cady (1982). *Analyzing Experimental Data by Regression*. California: Lifetime Learning Publications, p. 179.

Walking babies. Zelazo, Phillip R., Nancy Ann Zelazo and Sarah Kolb (1972). "Walking in the newborn, " *Science*, **176**, 314–315, via Larsen and Marx.

Percentage taking SAT. *WABF 1994*, p. 199.

Death penalty. Bedau, H. A. (1982). *The Death Penalty in America*, 3rd ed. New York: Oxford University Press, p. 217.

Topic 6

Random rectangles. Scheaffer, R. L., M. Gnanadesikan, A. Watkins and J. A. Witmer (1996). *Activity-Based Statistics: Student Guide*. New York: Springer-Verlag, p. 101.

Walking babies. (See Topic 5.)

Finger tapping. (See Topic 1.)

Kosslyn's imagery experiment. Kosslyn, S. M. (1980). *Image and Mind*. Cambridge, MA: Harvard University Press.

Topic 10

Penny spinning. Scheaffer, R. L., M. Gnanadesikan, A. Watkins and J. A. Witmer (1996). *Activity-Based Statistics: Student Guide*. New York: Springer-Verlag, p. 88.

Woburn wells. (See video listings)

Oral contraceptives. Sartwell, P. E., A. T. Masi, F. G. Arthes, G. R. Greene and H. E. Smith (1969). "Thromboembolism and oral contraceptives: an epidemiologic case control study," *Am. J. Epidemiol.*, **90**, 365.

Nutrasweet. (See video listing)

Ultracell. (See video listing)

Rowboat and ocean liner. Box, G. E. P. (1953). "Non-normality and tests on variances," *Biometrika*, **40**, 318–335.

Topic 11

Bartlett's quotations. Bartlett, John (1938). *Familiar Quotations*, 11th edition, Christopher Morley, ed. Boston: Little, Brown and Company.

Electrical insulation. Shewhart, W. A. (1986). *Statistical Method from the Viewpoint of Quality Control*. New York: Dover Publications, Inc., p. 90.

Phone listings. Springfield (MA) Telephone Directory, 1994–1995.

Topic 12

SAT scores. (See Topic 5.)

Percentage taking. (See Topic 5.)

Urban population. (See Topic 5.)

Dormitory population. (See Topic 5.)

Ice cream consumption. Koteswara Rao Kadiyala (1970). "Testing for the independence of regression disturbances," *Econometrica*, **38**, 97–117.

Anscombe plots. (See Topic 5.)

Cortisol and psychosis. Rothschild, A. J., A. F. Schatzberg, A. H. Rosenbaum, J. B. Stahl and J. O. Cole (1982). "The dexamethesone suppression test as a discriminator among subtypes of psychotic patients," *British Journal of Psychiatry*, **141**, 471–474.

Dopamine and schizophrenia. (See Topic 5.)

Walking Babies. (See Topic 5.)

Race and the death penalty. (See Topic 5.)

Women's salaries. Bellas, M. And B. F. Reskin. "On Comparable Worth," *Academe*, **80**, No. 5, p. 83–85.

Finger tapping. (See Topic 1.)

Pigs and vitamins. Iowa Agricultural Experiment Station, Animal Husbandry Swine Nutrition Experiment No. 577 (1952), via Snedecor and Cochran, p. 345.

Topic 13

Darwin's plants. Darwin, Charles (1876). *The Effect of Cross- and Self-fertilization in the Vegetable Kingdom,* 2nd ed., London: John Murray.

Hospital carpets. Walter, W. G. and A. Stober (1968). "Microbial air sampling in a carpeted hospital," *Journal of Environmental Health,* **30,** 405, via Larsen and Marx.

Walking babies. (See Topic 5.)

Gummy bears. Cobb, George W., "Gummy bears in space: factorial designs and interaction," in Scheaffer, R. L., M. Gnanadesikan, A. Watkins and J. A. Witmer (1996). *Activity-Based Statistics: Student Guide.* New York: Springer-Verlag, p. 188.

Mothers' stories. Werner, M., J. B. Stabenau and W. Pollin (1970). "TAT method for the differentiation of the families of schizophrenics, delinquents, and normals," *Journal of Abnormal Psychology* **75,** 139–145.

B. DATA SOURCES FOR THE SELF-TEST QUESTIONS*

Topic 1

Jumping bugs. Schmidt-Nielsen, Knut (1983). *Animal Physiology,* 3rd. ed. Cambridge: Cambridge University Press. p. 429.

Videotape ratings. *Consumer Reports,* November 1994, p.706.

Applied Materials. *USA Today,* 8/17/95, p. 3B.

Radios and mental defectives. Tufte, E. R. (1974). *Data Analysis for Politics and Policy.* New Jersey: Prentice-Hall, Inc., p. 89–90.

Stock Prices. *Business Week* no. 3432, 7/10/95, pp. 64ff.

Percentage below poverty. *AA,* p. 458.

Big Mac economics. Cook, R. D. And S. Weisberg (1994). *An Introduction to Regression Graphics.* New York: Wiley-Interscience Publication, p. 48.

Industry code. *Forbes,* v. 153, no. 9, 4/25/94, pp. 244ff.

Martin v. Westvaco. Robert A. Martin v. Envelope Division of WESTVACO Corporation, C.A. No. 93–10406–Z.

Discrimination claims. *USA Today,* 8/15/95.

Ali's fights. Merserole, M., ed. *Sports Almanac 1996.* Boston: Houghton Mifflin Company, p. 900.

Smart cards. *USA Today,* 8/17/95.

Counterfeit bills. *USA Today,* 8/16/95.

Stolen vehicles. *USA Today,* 8/15/95.

Baseball on radio. *USA Today,* 8/16/95.

---

*Sources that are listed in the text of the questions are not repeated here.

Topic 2

US travellers. *USA Today*, 8/18/95, p. 7D.

Income distribution. Phillips, K. (1993). *Boiling Point: Democrats, Republicans and the Decline of Middle-Class Prosperity*. New York: 1st Harper Perennial ed., Appendix A.

Spouse murder. Pie charts by Associated Press, via *Daily Hampshire Gazette*, 10/13/95, based on data from U.S. Department of Justice.

Best paid executives. Phillips, K. (1990). *The Politics of Rich and Poor*. New York: Random House, p. 179.

Top 15 stock funds. *USA Today*, 8/18/95.

States' dates of entry. *WABF 1996*, p. 655–681.

Justices' ages. *WABF 1996*, p. 92.

Largest floods. *WABF 1996*, p. 268.

Top 100 corporations. (See List B, Topic 1)

People per lawyer. *AA*, p. 192.

Baseball attendance. *USA Today*, 8/11/95, p. 3C.

Corporate yields. (See List B, Topic 1)

Death rates. *AA*, p. 79.

Land areas. *AA*, p. 205.

Highest points. *AA*, p. 209.

Hibernating hamsters. Acampora, Kelly Ann (1978). *The Photoperiodic Effects of* Na^+, K^+-*Adenosinetriphophatase Activity in the Golden Hamster*, undergraduate honors thesis, Department of Biological Sciences, Mount Holyoke College.

Stock prices. (See List B, Topic 1.)

Trial lawyers. Phillips, K. (1990). *The Politics of Rich and Poor*. New York: Random House, p. 176.

High temperatures. *AA*, p. 183.

Obedience to authority. Milgram, Stanley (1975). *Obedience to Authority*. New York: Harper & Row, Publishers.

Executions. *Newsweek*, 8/7/95, p. 25.

Topic 3

Acceptance rates. *ABC*, p. 46–48 and p. 54–56.

Percent below poverty. (See List B, Topic 1.)

Parkfield earthquakes. (See video credits)

Lowest temperatures. *AA*, p. 183.

Corporate sales. (See List B, Topic 1)

Change in share price. (See List B, Topic 1)

Admissions yields. *ABC*, p. 46–48 and p. 54–56.

MIT and Cal Tech. *ABC*, p. 46–48 and p. 54–56.

Student/faculty ratios. *ABC*, p. 46–48 and p. 54–56.

Alumni giving. *ABC*, p. 46–48 and p. 54–56.
Criminal brains. Gould, S. J. (1981). *Mismeasure of Man*. New York: W. W. Norton & Company, p. 94–95.

Topic 4

Hershey bars. Gould, S. J. (1983). "Phyletic Size Decrease," *Hen's Teeth and Horse's Toes*. New York: W. W. Norton & Company, p. 315.
People per doctor. *Daily Hampshire Gazette*, 12/26/95.
Farm employees. *WABF 1996*, p. 137.
VCRs. *AA*, p. 551.
Sunspots. (See List A, Topic 4.)
Carbon dioxide at Mauna Loa.
     http://gds.esrin.esa.it/0xc06afc3d_0x002d08e
Airline revenues. *AA*, p. 246.
Rail revenues. *AA*, p. 246.
Total transportation revenues. *AA*, p. 246.
Airline employees. *AA*, p. 246.
Air pollution. *AA*, p. 233.
Cost of a postcard. *AA*, p. 550.

Topic 5

Morton's skulls. Gould, S. J. (1981). *Mismeasure of Man*. New York: W. W. Norton & Company, p. 54.
Midterm elections. Tufte, E. R. (1974). *Data Analysis for Political and Policy*. New Jersey: Prentice-Hall, Inc., p. 142.
Student/faculty ratio. *ABC*, p. 46–48 and p. 54–56.
Mothers' stories. (See List A, Topic 13.)
Cost of a hospital room. *AA*, p. 113.
Number of doctors. *AA*, p. 109.
People per hospital. *AA*, p. 22 and p. 113.
Number of dentists. *AA*, p. 109.
Percentage of income spent on health. *AA*, p. 433.
Per capita income. *AA*, p. 431.
HS graduation rate. *AA*, p. 162.
Number on medicare. *AA*, p. 102.
Number on medicaid. *AA*, p. 102.
People per doctor. *AA*, p. 22 and p. 119.
People per nurse. *AA*, p. 22 and p. 119.
Percent 65 or older. *AA*, p. 38.
Number of hospitals. *AA*, p. 113.
Spending per student. *ABC*, p. 46–48 and p. 54–56.
SAT percentiles. *ABC*, p. 46–48 and p. 54–56.
Acceptance rate. *ABC*, p. 46–48 and p. 54–56.

Graduation rate. *ABC*, p. 46–48 and p. 54–56.

Retention rate. *ABC*, p. 46–48 and p. 54–56.

Alumni giving. *ABC*, p. 46–48 and p. 54–56.

Admissions yield. *ABC*, p. 46–48 and p. 54–56.

Spending per student. *ABC*, p. 46–48 and p. 54–56.

Percentage in the top 10th. *ABC*, p. 46–48 and p. 54–56.

Hibernating hamsters. (See List B, Topic 1.)

Urban percentage. *AA*, p. 22 and p. 27.

Vehicle-related death rates. *AA*, p. 609.

Minimum age for a license. *WABF 1996*, p. 211.

Cars per person. *WABF 1996*, p. 211.

Population density. *AA*, p. 22 and p. 205.

Median family income. *WABF 1996*, p. 383.

Women's salaries. (See List A, Topic 12.)

Number of MDs. *AA*, p. 109.

Number of deaths. *AA*, p. 79.

Walking babies. (See List A, Topic 5.)

Visual and verbal. Brooks, L. R. (1968). "Spatial and verbal components of the act of recall," *Canadian Journal of Psychology* **22**, 349–368.

Leafhopper survival. (See List A, Topic 5.)

Hospital carpets. (See List A, Topic 13.)

Coporate assets. (See List B, Topic 1.)

Race bias at the Fed. Finkelstein, M. O. And B. Levin (1990). *Statistics for Lawyers*. New York: Springer-Verlag, p. 170.

Oral contraceptives. (See List A, Topic 10.)

Portacaval shunt. Grace, N. D., H. Muench, and T. C. Chalmers (1966). "The present status of shunts for portal hypertension in cirrhosis," *J. Gastroenterology*, Vol. 50, 646–691

Smith College infirmary. Hodge, Mary Beth (1976). Undergraduate honors thesis, Department of Mathematics, Smith College.

Factors related to the death penalty. Finkelstein, M. O. And B. Levin (1990). *Statistics for Lawyers*. New York: Springer-Verlag, p. 190.

Factionalism and wealth. Boyer, P. and S. Nissenbaum (1972). *Salem-Village Witchcraft: A Documentary Record of Local Conflict in Colonial New England*. California: Wadsworth Publishing Company.

Hiring at the University of Texas. Finkelstein, M. O. and B. Levin (1990). *Statistics for Lawyers*. New York: Springer-Verlag, p. 244.

Topic 6
Gastric freezing. Wangensteen, O. H. (1962). *Journal of the American Medical Association.*, **180**, 439–444.

Hospital carpets. (See List A, Topic 13.)

Remembering words. Data from student lab project, Department of Psychology, Mount Holyoke College.

Finger tapping. (See List A, Topic 1.)

Launching bears. (See List A, Topic 13.)

Topic 7
Welfare mothers. Harper's Index, *Harpers*, March 1995, p. 11.

Underwear. Harper's Index, *Harpers*, February 1995, p. 13.

Swiss Army knives. Harper's Index, *Harpers*, February 1995, p. 13.

Ads on bills. Harper's Index, *Harpers*, November 1995, p. 11.

Topic 8
ACT scores. Personal communication to Jonathan D. Cryer from the Registrar, University of Iowa.

Topic 9
Zinc in fish. *Environmental Monitoring and Assessment*, 1993

Nicotine patches. Hurt, R.L. et al, "Nicotine patch therapy for smoking cessation combined with physician advice and nurse follow-up," *Journal of the American Medical Association.*, Feb. 23, 1994, **271**, No. 8, pp. 595–600.

Mozart. Rauscher, F.H. et al, "Music and spatial task performance," *Nature*, **365**, Oct. 14, 1993, p. 611.

Topic 10
Translation errors. OmniLingua: personal communication 1996, to Jonathan D. Cryer from Mark Sellergren, CEO, Omni-Lingua, Inc., Cedar Rapids, Iowa.

Body dehydration. Dale, G., J. A. Fleetwood, A. Weddell, R. D. Ellis and J. R. C. Sainsbury (1987). "Beta endorphin: a factor in 'fun run' collapse?" *British Medical Journal*, **294**, 1004, via Hand.

Topic 12
Counting manatees. Rathbun, Galen B., "Fixed-wing versus heliocopter surveys of manatees," *Marine Mammal Science* 4(1), 1988, pp. 71–75.

Ice cream consumption. (See List A, Topic 12)

Jellyfish. Lund, A. D. and McNeil, D. R. (1991). *Computer-Interactive Data Analysis*. Chichester: John Wiley & Sons, p. 308, via Hand.

Hardness and density. Williams, E. J. (1959). *Regression Analysis*. New York: John Wiley & Sons, Inc., p. 43, via Hand.

Bubblebaths. Student projects in classes of Jonathan D. Cryer, University of Iowa.

Hardness of pencil leads. Student project reported in Vardeman, Stephen (1994). *Statistics for Engineering Problem Solving*, PWS-Publishing, 1994, p. 433.

Brands of pencil lead. Student project in the class of Russell Lenth, University of Iowa.

Brands of popcorn. Student project in the class of Russell Lenth, University of Iowa.

Gasoline mileage. *Consumer Reports*, various issues 1987–1988, via Jonathan D. Cryer and Robert B. Miller (1991). *Statistics for Business: Data Analysis and Modelling*. Boston: PWS-Kent Publishing Company.

Glue strength. Vardeman, Stephen (1994). *Statistics for Engineering Problem Solving*, PWS-Publishing, 1994, p. 514.

Topic 13

Bee stings. Free, J. B. "The Stimuli Releasing the Stinging Response of Honeybees," *Animal Behavior*, 9, 1961, p. 193–196.

Presurgical stress. Hoaglin, D. C., F. Mosteller, and J. W. Tukey (1985). *Exploring Data Tables. Trends and Shapes*. New York: John Wiley & Sons, p. 420, via Hand.

Smoking and chickenpox. Ellis, M. E., K. R. Neal and A. K. Webb (1987). "Is smoking a risk factor for pneumonia in patients with chickenpox," *British Medical Journal*, **294**, 1002, via Hand.

Cortisol and schizophrenia. (See List B, Topic 12)

Cortisol and psychotics. (See List B, Topic 12)

Frequency of alpha waves. Gendreau, Paul, et al. (1957). "Changes in EEG alpha frequency and evoked response latency during solitary confinement," *Journal of Abnormal Psychology*, **79**, 54–59, via Larsen and Marx.

Hospital carpets. (See List A, Topic 13.)

Solitary confinement. (See frequency of alpha waves, above.)

Cholesterol and personality. Selvin, S. (1991). *Statistical Analysis of Epidemiological Data*. New York: Oxford University Press, Table 2.1, via Hand.

Finger tapping. Draper, N. R. and H. Smith (1981). *Applied Regression Analysis, 2nd edition*. New York: John Wiley & Sons, p. 425, via Hand.

Diabetes and thromboglobulin. Van Oost, B. A., B. Veldhayzen, A. P. M., Timmermans, and J. J. Sixma (1983) "Increased urinary β-thromboglobulin excretion in diabetes assayed with a modified RIA kit-technique," *Thrombosis and Haemostasis*, **9**, 18–20, via Hand.

Darwin's plants. (See List A, Topic 13.)

Walking babies. (See List A, Topic 5.)

Sleeping shrews. Berger, R. J. and J. M. Walker (1972). "A polygraphic study of sleep in the tree shrew," *Brain, Behavior and Evolution*, **5**, 62, via Larsen and Marx.

C. DATA SOURCES FOR THE SOLVED PROBLEMS*

Topic 1

2. Boyer, P. and S. Nissenbaum (1972). *Salem-Village Witchcraft: A Documentary Record of Local Conflict in Colonial New England*. California: Wadsworth Publishing Company.

5. Baseball attendance. (See List B, Topic 2.)

7. Baseball on radio. (See List B, Topic 1.)

Topic 2

15. Baseball standings. Merserole, M., ed. *Sports Almanac 1996*. Boston: Houghton Mifflin Company, p. 83.

16. Osmoregulation. Department of Biological Sciences, Mount Holyoke College.

18. Stock markets in Europe. *Worth*, July/August 1994, p.26.

38. Mantle's averages. *USA Today*, 8/14/95.

39. Salty chips. *Consumer Reports*, August, 1995.

40–44. Women's salaries. (See List A, Topic 12.)

45. Old airplanes. *USA Today*, 7/5/95.

46. Radioactive twins. Camner, Per, and Philipson, Klas (1973). "Urban factor and tracheobronchial clearance," *Archives of Environmental Health*, **27**, 82, via Larsen and Marx.

49, 51. Accused witches. Boyer, P. and S. Nissenbaum (1972). *Salem-Village Witchcraft: A Documentary Record of Local Conflict in Colonial New England*. California: Wadsworth Publishing Company.

52, 53. Hiring patterns. (See Martin v. Westvaco, List B, Topic 1.)

54. Exports. *AA*, p. 796.

55. Farms. *AA*, p. 644.

56. Temperature extremes. *WABF 1994* , p.166.

57. Hazardous waste sites. (See List A, Topic 3.)

58. Drivers. *AA*, .p. 606.

59. Wisconsin. (See List B, Topic 6.)

*Sources that are listed in the text of the problems are not repeated here.

Topic 3
1. Parkfield earthquakes. (See video listing)
2. Boston marathon times. Merserole, M., ed. *Sports Almanac 1996*. Boston: Houghton Mifflin Company, p. 638.

Topic 4
2. Population of Florida. US Bureau of the Census via Witmer, J. A. (1992). *Data Analysis: An Introduction*. New Jersey: Prentice Hall.
4. Larcenies. US Department of Justice, *National Crime Survey Report*, via Witmer, J. A. (1992). *Data Analysis: An Introduction*. New Jersey: Prentice Hall.
7. Math PhDs. Fulton, John D. (1994). "1994 AMS-IMS-MAA Annual Survey," *Notices of the American Mathematical Society*, **41**, pp. 1121–1136 .

Topic 5
1. Lung cancer patients. Moses, L. E. (1986). *Think and Explain with Statistics*. Massachusetts: Addison-Wesley Publishing Company, p. 159.
2. Hibernating hamsters. (See List B, Topic 1.)

Topic 9
12. Screening for hepatitis. Prince, A. M., and R. K. Gershon. "The use of serum enzyme determinations to detect anicteric hepatitis," *Transfusion*, 5:120, 1965.

Topic 10
12. Dehydration. (See List B, Topic 11.)

Topic 12
15. Prisoners in solitary. (See List B, Topic 13.)
28. Moon phases and mental health. Olvin, J. F. (1943). "Moonlight and nervous disorders," *American Journal of Psychiatry*, **99**, 578–84, via Larsen and Marx.
29. Pigs and antibiotics. (See List A, Topic 12.)
32. Sleeping shrews. (See List B, Topic 13.)
35. Gasoline consumption. (See List B, Topic 12.)
43. Suess verses. Based on a project done by Anne Wilson, class of 1989, Oberlin College.
    Walking babies. (See List A, Topic 5.)

Topic 13

8. Fish species and location. Based on a project done by Dan Rogers and Michael Heithaus, class of 1995, Oberlin College.

9. Cigarette smoking and gender. Based on a project done by Karina Quon, class of 1997, Oberlin College.